STUDENT UNIT GUIDE

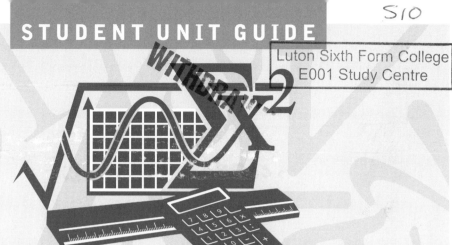

A2 Mathematics
UNIT 4723

Module 4723: Core Mathematics 3

Lawrence Jarrett

Philip Allan Updates, an imprint of Hodder Education, an Hachette UK company,
Market Place, Deddington, Oxfordshire OX15 0SE

Orders

Bookpoint Ltd, 130 Milton Park, Abingdon, Oxfordshire OX14 4SB
tel: 01235 827720
fax: 01235 400454
e-mail: uk.orders@bookpoint.co.uk
Lines are open 9.00 a.m.–5.00 p.m., Monday to Saturday, with a 24-hour message
answering service. You can also order through the Philip Allan Updates website:
www.philipallan.co.uk

© Philip Allan Updates 2009

ISBN 978-1-84489-576-2

First printed 2009
Impression number 5 4 3 2 1
Year 2013 2012 2011 2010 2009

This guide has been written specifically to support students preparing for the
OCR A2 Mathematics Unit 4723 examination. The content has been neither
approved nor endorsed by OCR and remains the sole responsibility of the author.

Typeset by Pantek Arts Ltd, Maidstone
Printed by MPG Books, Bodmin

Hachette Livre UK's policy is to use papers that are natural, renewable and
recyclable products and made from wood grown in sustainable forests. The
logging and manufacturing processes are expected to conform to the
environmental regulations of the country of origin.

8000S140

WITHDRAWN

A2 Mathematics

Contents

Introduction

■ ■ ■

Content Guidance

General Matters

Questions and Answers

Introduction

About this guide

This unit guide is one of a series covering the OCR specification for AS and A-level mathematics. It offers advice for the effective study of **A2 Module 4723: Core Mathematics 3 (C3)**. The guide has four sections:

- **Introduction** — this includes some general points about the specification, about how to revise and about how to tackle the unit examination itself.
- **Content Guidance** — this covers the topic areas in the specification, highlights the key points and provides worked examples. You should attempt to answer each of these worked examples by yourself before checking the given solution and reading the accompanying notes. Each topic is followed by a brief exercise, to give you the chance to consolidate your understanding.
- **General Matters** — this section discusses the appropriate and efficient use of calculators and the formulae booklet; it also provides information about synoptic assessment.
- **Questions and Answers** — this section contains three specimen papers. Hints and suggestions are provided, together with solutions to the questions. Each solution is followed by an examiner's comment. The answers to the exercises that occur throughout the book are given at the end of the section.

Core Mathematics 3 is an A2 module. In the examination, some of the questions will be relatively straightforward and routine, while others will contain more challenging aspects. If you have a thorough understanding of all the topic areas, you will be able to approach the examination with confidence. Of course, hard work is needed; mathematics is best revised by actually doing mathematics, which is why this guide contains plenty of questions for you to attempt.

The specification

The detailed specification for C3 can be found at **www.ocr.org.uk**. Mathematics cannot be divided into discrete parts, so you will need knowledge learned earlier in your study of mathematics. Indeed, the C3 specification states that 'Knowledge of the specification content of Modules C1 and C2 is assumed, and candidates may be required to demonstrate such knowledge in answering questions in Unit C3.' This does not mean that a question in the C3 examination will necessarily be based on a topic from an earlier module, but it does indicate that when solving a C3 question, you may well need to recall a result or technique from C1 or C2.

How to answer questions

Mathematics is a discipline in which accuracy, precision and care are vital. First, read the question carefully — and then re-read it. Adopt a careful and thoughtful approach to setting out your solution; neat and tidy work helps to prevent careless slips. Always double-check that you have written down details from the question accurately in your solution. If you carelessly change the inequality $|3x - 8| < |2x - 1|$ to $|3x - 8| < |2x + 1|$ or the integral $\int \sqrt[3]{6x + 5}\,dx$ to $\int \sqrt{6x+5}\,dx$, you will obviously get the wrong answer. Indeed, such an error may prevent you from making any progress with a solution.

Once you think you have completed the solution, go back to the question and check that you really have answered what was asked. Make sure that you have answered all the parts of a question. Don't be so pleased about having answered parts (i) and (ii) that you overlook the fact that the question also has a part (iii).

Remember that someone is going to look at your solutions. It is in your interests to present your work clearly so that the examiner has an easy job in reading and under-standing all aspects of your solutions.

Sketch graphs

A question may require you to sketch a graph. Precise plotting of points is not neces-sary, and there is no need to use graph paper. The sketch should be drawn neatly in the answer booklet, showing the basic shape and essential characteristics of the curve involved.

Part questions

Some questions are divided into parts. If the parts are labelled (a), (b) and (c), then they are quite separate and the answer you obtain in part (a), for example, will be of no relevance to part (b). If the parts are labelled (i), (ii) and (iii), then they may be linked and it is possible that what you have done in part (i) will have some bearing on what you are to do in part (ii).

Look out for words such as 'Hence' and 'Deduce'. These signal that the part of the question using such a word definitely follows on from a previous part and that an earlier result will probably be needed.

Exact answers

A question may ask for an exact answer to be given. This means that the method you adopt must be exact throughout. Decimal approximations are not acceptable in such cases, no matter how many decimal places you offer; for example, $\frac{1}{7}\pi\sqrt{3}$ is an exact number, but both 0.247π and 0.777 are merely approximations of its value. An exact answer could be requested because the examiner wishes to assess a particular method of solution. Therefore, if you resort to providing decimal values, you are likely to lose several marks.

Significant figures

Unless the question stipulates otherwise, a numerical answer that is not exact should be given correct to three significant figures. This does not mean that you should do all your working with only three significant figures. In fact, calculations must be carried out using greater accuracy — use your calculator as much as possible, and retain as many digits as you can, at intermediate steps of working — so that, at the end of the solution, you can confidently judge what the true three-significant-figure version of the answer ought to be.

Proofs

A question might ask you to prove or show a result that is quoted in the question. In this case, your solution must be particularly detailed so that you are able to convince the examiner that you know exactly what you are doing.

Sometimes a result is given in the question because it is to be used in a subsequent part of the question. Even if you are unable to prove the given result, you are still entitled — and expected — to use it in answering subsequent parts.

The unit examination

The C3 examination lasts for 1 hour 30 minutes and will consist of around nine questions. You have to attempt *all* the questions. This is an A2 module, so although some of the questions will be relatively routine and straightforward, there will also be aspects that are more challenging and which are designed to test the depth of your mathematical understanding.

The total for the paper is 72 marks. Questions with lower numbers of marks will be towards the beginning of the paper. You should answer the questions in numerical order, leaving a question unfinished only if you are stuck.

The number of marks allocated to a question is a guide to how much work is expected. Do not expect to answer a question worth 7 marks in just two lines; on the other hand, a question worth 2 marks should certainly not take a whole page for the answer. Allowing time for thinking and checking, a total of 72 marks in an examination lasting 90 minutes indicates an approximate rate of 'a mark a minute'.

- Make sure that you arrive at the examination in good time, so that you are calm and fully prepared for the start of the paper.
- Read each question carefully.
- Check that you have transferred details from the question, such as figures and equations, correctly into your solution.
- Tackle each question (including the early, easier ones) in a steady, thoughtful manner — a casual, rushed approach leads to carelessness and to unnecessary loss of marks.

How to use this guide

You will do best if you have a thorough, planned revision programme. Start your revision in good time and set yourself manageable amounts to do each day. This guide contains plenty of questions so that you can practise what you have to do in the examination itself, i.e. answer questions. You do not have to work through this guide in order; a better plan might be to mix topics from functions, calculus and trigonometry. Five of the exercises — Exercise 9, Exercise 18, Exercise 25, Exercise 32 and Exercise 35 — consist of more demanding questions. You might like to challenge yourself with these at an appropriate stage.

Once you are confident with all the topics of C3, set aside a period of 90 minutes when you can work without interruption and try one of the specimen papers. The Questions and Answers section has further advice on how to approach these papers in the most effective and beneficial way.

Enjoy working steadily and methodically through this guide. I hope you will find the guide helpful and offer my best wishes for the examination.

Content
Guidance

The content of **A2 Module 4723: Core Mathematics 3 (C3)** consists of five main topic areas — functions, differentiation, integration, trigonometry and solution of equations.

Algebra is an important feature of all A-level mathematics modules; in this module, the focus is on functions. The theory of functions is developed, and new functions — such as modulus functions, exponential functions and logarithmic functions — are introduced and explored.

Differentiation was introduced in module C1, and integration in C2. This module develops both areas extensively, presenting new techniques that enable more complicated functions to be analysed.

In trigonometry, the secant, cosecant and cotangent functions are introduced, as well as the inverse functions of sine, cosine and tangent. A number of identities are proved; application of these identities will allow you to manipulate more complicated trigonometric expressions and to solve more involved trigonometric equations.

Not all equations can be solved by purely algebraic means. A short final section deals with a particular method of finding approximate roots to equations by using a numerical approach.

The solutions to the worked examples contain explanatory notes, which are preceded by the icon *n*.

Functions

Revision

In earlier modules, the notation f(x) was used occasionally to refer to a function of a variable x. For example, if $f(x) = x^2 + 3x$, then $f(5) = 5^2 + 3 \times 5 = 40$. In this module, functions are studied more extensively.

Domain and range

A function is not fully defined until its **domain** is specified, i.e. until we know the set of values of x that are permitted to be substituted into the function's formula. For example, consider the function f defined by $f(x) = x^2 + 4$ for $x \geq 0$. The domain of f consists of values of x that are non-negative. The **range** of a function is the set of values which result from substituting all the values of the domain into the formula. In this case, the range of f is $f(x) \geq 4$. The corresponding graph of $y = f(x)$ is shown below, with the domain and range indicated.

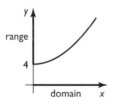

Worked example 1

The function f is defined by $f(x) = x^2 - 4x + 1$ with domain consisting of all real numbers. Determine the range of f.

Solution

11 It will be helpful to complete the square for this quadratic expression and sketch its graph.

$f(x) = x^2 - 4x + 1 = (x - 2)^2 - 3$

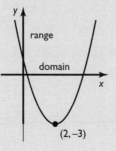

The vertex of the parabola is (2, –3), so the range is $f(x) \geq -3$

Worked example 2

The function g is defined by $g(x) = \sqrt{x}$ for the domain $4 \leq x \leq 9$. Sketch the graph of $y = g(x)$ and determine the range of g.

Solution

The domain is $4 \leq x \leq 9$, so the graph exists only for these values.

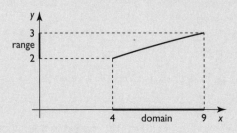

The range is the set of permitted y-values, i.e. the values from $\sqrt{4}$ to $\sqrt{9}$
Thus the range is $2 \leq y \leq 3$ or $2 \leq g(x) \leq 3$

Exercise 1

(1) The function f is defined by $f(x) = x^2 - 5$ for all real values of x. Find the range of f.

(2) The function g is defined by $g(x) = \sin x$ for $0° < x < 360°$. State the range of g.

(3) The function f is defined by $f(x) = 2x^2 + 4x + 11$ for all real values of x. Find the range of f.

Composition of functions

Consider the functions $f(x) = 5x - 3$ and $g(x) = x^2 + 1$, both defined for all real values of x.

Suppose that, first, f maps 2 to 7, i.e. $f(2) = 7$; and, then, g maps 7 to 50, i.e. $g(7) = 50$. These two steps can be written in combination as $gf(2) = g(7) = 50$.

Pay attention to the order in which the composition of the two functions is written: the second function applied is written to the *left* of the first function applied. The order is important because $fg(2)$ gives a different answer: $fg(2) = f(5) = 22$.

Worked example 1

Given $f(x) = 3 - 5x$ and $g(x) = x^3 - 6$, where f and g both have domains consisting of all real numbers, find $fg(3)$.

Solution

⚠ Note that fg means that g is applied first and then f.
$fg(3) = f(3^3 - 6) = f(21) = 3 - 5 \times 21 = -102$

content guidance

Worked example 2

Functions f and g are defined by f: $x \mapsto 4x + 7$ for $x \in \mathbb{R}$ and g: $x \mapsto (x + 2)^2$ for $x \in \mathbb{R}$.
Find

(i) ff   **(ii)** gf

Solution

> $\boxed{n}$ f: $x \mapsto 4x + 7$ is an alternative way of presenting the function $f(x) = 4x + 7$.
> ff means that the function f is applied to x and then f is applied to the result,
> i.e. f is applied twice in succession. $\mathbb{R}$ represents the set of real numbers, so
> $x \in \mathbb{R}$ just says that the domain consists of all real numbers.

(i) $ff(x) = f(4x + 7) = 4(4x + 7) + 7 = 16x + 35$
so $ff : x \mapsto 16x + 35$

(ii) $gf(x) = g(4x + 7) = ((4x + 7) + 2)^2 = (4x + 9)^2$
so $gf : x \mapsto (4x + 9)^2$

Exercise 2

(1) The functions f and g are defined for all real values of x by $f(x) = 7x - 5$ and $g(x) = 2x + 6$.
Find

(i) fg(−1)   **(ii)** gf(x)

(2) The function h is defined for $x \geq 8$ by $h(x) = \sqrt{x - 8}$. The function f is defined for $x \in \mathbb{R}$ by
$f(x) = 2^x$. Find

(i) hf(5)   **(ii)** fh(x)

Inverse functions

A function f is **one-one** (or **1–1**) if each value of its range is obtained from **only one**
value of the domain. The graph of a 1–1 function is such that a horizontal line $y = k$,
where k is any value of the range, will meet the graph exactly once. Of the following
graphs, two represent 1–1 functions and two represent functions that are not 1–1. Can
you tell which is which?

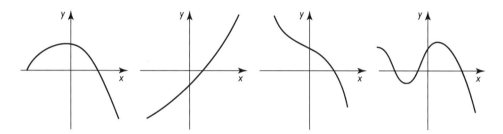

If a function f is 1–1, then its inverse f^{-1} exists. If the function is such that $f(3) = 8$, then
its inverse f^{-1} maps 8 to 3, i.e. $f^{-1}(8) = 3$.

Worked example 1

Given that f is defined by $f(x) = x^3 + 4$ for $x \in \mathbb{R}$, show that f is 1–1 and evaluate $f^{-1}(12)$.

Solution

n We can demonstrate that f is 1–1 by sketching the graph of $y = f(x)$.

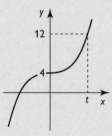

Each value of the range (which is the whole y-axis) is associated with exactly one x-value, i.e. any straight line parallel to the x-axis meets the curve exactly once.

If $f^{-1}(12) = t$, then, as shown in the graph, $f(t) = 12$

i.e. $t^3 + 4 = 12$, giving $t = 2$

Hence $f^{-1}(12) = 2$

n f maps 2 to 12, whereas the inverse f^{-1} maps 12 to 2.

Worked example 2

Given $f(x) = 4x - 9$, find $f^{-1}(x)$.

Solution

f maps x to y, where $y = 4x - 9$

f^{-1} maps y to x, where x is an expression in terms of y

Rearranging $y = 4x - 9$ to make x the subject, we get $x = \frac{1}{4}(y + 9)$

So f^{-1} maps y to $\frac{1}{4}(y + 9)$, i.e. $f^{-1}(y) = \frac{1}{4}(y + 9)$

or, changing the name of the variable back to x, $f^{-1}(x) = \frac{1}{4}(x + 9)$

For any function with an inverse, there is a geometrical relationship between the graph of $y = f(x)$ and the graph of $y = f^{-1}(x)$. Each is the reflection of the other in the straight line $y = x$.

For f and f^{-1} in Worked example 2 above, the respective graphs are shown together with the line of symmetry $y = x$.

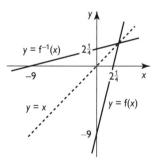

Exercise 3

(1) The function f is defined by $f(x) = \frac{1}{2}x^3 + 5$ for $x \in \mathbb{R}$. Find an expression for $f^{-1}(x)$.

(2) Given $g(x) = \sqrt{2x + 7}$, find $g^{-1}(5)$.

(3) The function f is defined by $f(x) = 2^x$ for all real values of x. Sketch, on the same diagram, the graphs of $y = f(x)$ and $y = f^{-1}(x)$.

Modulus functions

The modulus of x is written $|x|$ and means the non-negative value of x, i.e. the **magnitude** of x. So $|8| = 8$, $|3.7| = 3.7$, $|-4| = 4$, $\left|-6\frac{1}{2}\right| = 6\frac{1}{2}$ and $|\sin 270°| = 1$.

The effect of the modulus sign is to replace a negative value by its corresponding positive value. Observe, in the following particular cases, how the graphs of $y = f(x)$ and $y = |f(x)|$ are related.

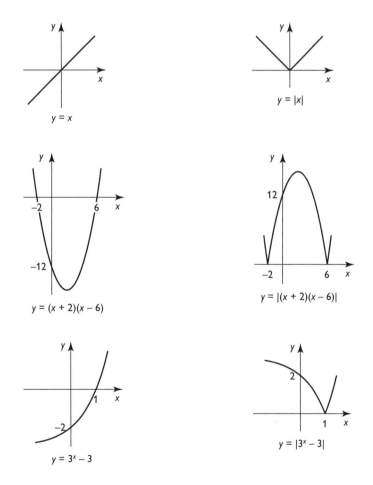

$y = x$

$y = |x|$

$y = (x + 2)(x - 6)$

$y = |(x + 2)(x - 6)|$

$y = 3^x - 3$

$y = |3^x - 3|$

In each case, that part of the $y = f(x)$ graph which lies below the x-axis is reflected in the x-axis.

Exercise 4

(1) Sketch the graph of $y = |2x - 4|$.

(2) Sketch the graph of $y = |x^2 - 9|$.

(3) Sketch, for $0° \leq x \leq 360°$, the graph of $y = |\sin x|$.

Modulus equations

There are two possible methods for solving an equation such as $|3x - 1| = 7$. The first method involves using the definition of modulus to form two linear equations. Since $|3x - 1| = 7$, it could be that $3x - 1 = 7$ or that $3x - 1 = -7$. Solving each of these two linear equations gives $x = \frac{8}{3}$ or $x = -2$.

The second method involves squaring both sides of the equation, which leads to $(3x - 1)^2 = 49$. Check that solving this quadratic equation yields the same two roots.

Worked example

Solve the equation $|5x + 2| = |2x - 16|$.

Solution

Method 1 (forming two linear equations):

$5x + 2 = \pm(2x - 16)$

that is, $5x + 2 = +(2x - 16)$ or $5x + 2 = -(2x - 16)$

leading to $3x = -18$ or $7x = 14$

Hence the two answers are $x = -6$ and $x = 2$

Method 2 (squaring both sides):

$(5x + 2)^2 = (2x - 16)^2$

giving $25x^2 + 20x + 4 = 4x^2 - 64x + 256$

which leads to $21x^2 + 84x - 252 = 0$ and (upon dividing by 21) $x^2 + 4x - 12 = 0$

Hence $(x + 6)(x - 2) = 0$

and so $x = -6$ or $x = 2$

11 Remember that each of the answers obtained can be checked by substituting it in the original equation. For example, $x = 2$ is correct because $|10 + 2| = |4 - 16|$ holds true. Either of the two solution methods is acceptable in the C3 examination; choose the one with which you feel more comfortable.

Exercise 5

(1) Solve $|2x + 5| = 11$.

(2) Solve $|x - 4| = |x + 8|$.

(3) Solve $|3x - 2| = |2x - 7|$.

Modulus inequalities

Worked example

Solve the inequality $|2x - 5| \leq |x + 2|$.

Solution

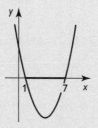 Various methods are possible for solving such inequalities, but probably the most direct one is to square both sides.

Squaring gives $(2x - 5)^2 \leq (x + 2)^2$

Expanding, $4x^2 - 20x + 25 \leq x^2 + 4x + 4$

Simplifying, $3x^2 - 24x + 21 \leq 0$ and hence $x^2 - 8x + 7 \leq 0$

Factorising, $(x - 1)(x - 7) \leq 0$

Checking the graph of $y = (x - 1)(x - 7)$:

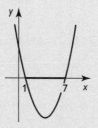

we see that the inequality is satisfied for $1 \leq x \leq 7$

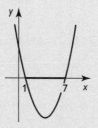 This method led to a quadratic inequality; the solution of this needs care and is best done by referring to the associated graph.

Exercise 6

(1) Solve the inequality $|4x - 1| > 3$.

(2) Solve the inequality $|2x + 1| < |2x - 5|$.

(3) Solve the inequality $|3x - 2| \geq |2x - 3|$.

Curve transformations

In module C1, you met translations, reflections and stretches of graphs. Here is a summary of these transformations:

- translation by a units in the positive x-direction: $y = f(x) \rightarrow y = f(x - a)$
- translation by a units in the positive y-direction: $y = f(x) \rightarrow y = f(x) + a$
- reflection in the x-axis: $y = f(x) \rightarrow y = -f(x)$
- reflection in the y-axis: $y = f(x) \rightarrow y = f(-x)$
- stretch by scale factor a in the x-direction: $y = f(x) \rightarrow y = f(\frac{x}{a})$
- stretch by scale factor a in the y-direction: $y = f(x) \rightarrow y = af(x)$

This module covers the composition of two or more of these transformations. In some cases, the order in which the transformations are applied will be vital.

Worked example 1

Give details of the curve transformations which map the graph of $y = x^3$ to the graph of $y = 4(x + 2)^3$.

Solution

Since x, the quantity being cubed, has been replaced by $x + 2$, one of the transformations is a translation in the negative x-direction by 2 units.

Since $(x + 2)^3$ has been multiplied by 4, the second transformation is a stretch in the y-direction by scale factor 4.

> **11** In this case, the order of the two transformations does not matter. The translation followed by the stretch gives $y = x^3 \rightarrow y = (x + 2)^3 \rightarrow y = 4(x + 2)^3$; the stretch followed by the translation gives $y = x^3 \rightarrow y = 4x^3 \rightarrow y = 4(x + 2)^3$.

Worked example 2

The curve $y = \dfrac{1}{x}$ undergoes the following transformations: translation by 3 units in the negative y-direction, followed by reflection in the y-axis and then a stretch by scale factor $\frac{1}{2}$ in the x-direction. Find the equation of the resulting curve.

Solution

Translation: $y = \dfrac{1}{x} \rightarrow y = \dfrac{1}{x} - 3$

Reflection: $y = \dfrac{1}{x} - 3 \rightarrow y = \dfrac{1}{(-x)} - 3 = -\dfrac{1}{x} - 3$

Stretch: $y = -\dfrac{1}{x} - 3 \rightarrow y = -\dfrac{1}{2x} - 3$

The resulting curve has equation $y = -\dfrac{1}{2x} - 3$

> **11** The stretch in the x-direction is the most awkward of the transformations. Note that the scale factor $\frac{1}{2}$ means that x should be replaced by $2x$ in the equation.

Exercise 7

(1) The graph of $y = |x|$ is first translated by 4 units in the positive x-direction and then stretched by scale factor $\frac{1}{2}$ in the y-direction. Give the resulting equation and sketch its graph.

(2) Give details of the transformations needed to transform the curve $y = x^2$ to the curve $y = -(2x + 5)^2$.

Exponentials and logarithms

You met functions such as 2^x and 5^{2x-1} in C2. In this module, a further exponential function plays a significant role, namely $\mathbf{e}^x$, where e is an irrational number whose value is approximately 2.7183. (e^x is an important function because its derivative happens to be exactly itself, whereas other exponential functions such as 2^x have more complicated derivatives.)

The graph of $y = e^x$ is as follows:

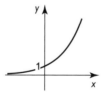

The function $f : x \mapsto e^x$ is a one-one function and its inverse is $f^{-1} : x \mapsto \ln x$, where ln denotes a logarithm with base e. Logarithms with base e are called **natural logarithms**. The general properties of logarithms apply to natural logarithms as well:

- $\ln 1 = 0$
- $\ln e = 1$
- $\ln(ab) = \ln a + \ln b$
- $\ln\left(\frac{a}{b}\right) = \ln a - \ln b$
- $\ln(a^p) = p \ln a$

The graph of $y = \ln x$ is as follows:

Note that the graphs of $y = e^x$ and $y = \ln x$ are reflections of each other in the line $y = x$. This is true because each is the inverse of the other.

Worked example 1

Solve the equation $e^{2x+1} = 7$.

Solution

By natural logarithms, $2x + 1 = \ln 7$

Hence $2x = \ln 7 - 1$ and therefore $x = \frac{1}{2} (\ln 7 - 1)$

n $\frac{1}{2} (\ln 7 - 1)$ is the exact value of x, and this is the form in which you are expected to present the answer. Its approximate decimal value, which you can check by using your calculator, is 0.473.

Worked example 2

Express $2 \ln a + 3 - 4 \ln b$ as a single logarithm.

Solution

$$2\ln a + 3 - 4\ln b = \ln a^2 + \ln e^3 - \ln b^4 = \ln\left(\frac{a^2 e^3}{b^4}\right)$$

 The logarithm properties are used here. Note that, because the base of natural logarithms is e, the number 3 can be written as $\ln e^3$.

Exercise 8

(1) Given that $2\ln(x + 5) = 3$, find x in terms of e.

(2) Express $3\ln 2 - \ln 3 + 5$ in the form $\ln p$.

(3) Solve the equation $7e^{3x} = 100$, giving the root correct to 3 significant figures.

General functions problems

Worked example

The function f is defined by $f(x) = x^2 - 4x + 12$ for $x \le k$. It is given that f is a 1–1 function. Find
(i) the greatest possible value of the constant k **(ii)** $f^{-1}(x)$

Solution

 The minimum point (vertex) of $y = x^2 - 4x + 12$ is significant.

(i) Completing the square, we get $f(x) = (x - 2)^2 + 8$
so the vertex of the parabola is at $(2, 8)$
The complete graph of $y = (x - 2)^2 + 8$ is as follows:

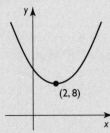

This is not the graph of a 1–1 function; but, if the domain is restricted to $x \le 2$, then the graph becomes

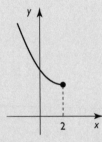

and this does show a 1–1 function.
Therefore, the greatest possible value of k is 2

> **(ii)** To find $f^{-1}(x)$, we need to rearrange $y = x^2 - 4x + 12$ to the form $x = ...$
> The completed square form $y = (x - 2)^2 + 8$ becomes $y - 8 = (x - 2)^2$
> hence $x - 2 = \pm\sqrt{y - 8}$, i.e. $x = 2 \pm \sqrt{y - 8}$
> The question now is whether to choose the positive or negative root, i.e.
> which formula is correct: $f^{-1}(x) = 2 + \sqrt{x - 8}$ or $f^{-1}(x) = 2 - \sqrt{x - 8}$?
> Testing a particular value will enable us to make the right choice.
> For example, f maps 0 to 12; so f^{-1} must map 12 to 0
> Substituting 12 for x in each of the two possible expressions for $f^{-1}(x)$, the
> first possibility gives 4 whereas the second gives 0
> Hence $f^{-1}(x) = 2 - \sqrt{x - 8}$

Exercise 9

(1) The function f is defined by $f(x) = 2^{3x-1}$ for all real values of x. Find $f^{-1}(x)$.

(2) Solve the equation $|2x - 5a| = |4x + 11a|$ for x.

(3) The function f is defined for all real values of x by $f(x) = 15 + 4x - 4x^2$. Express $f(x)$ in completed square form and hence find the range of f.

(4) Solve the inequality $|3x - 7| < |2x + 1|$. Illustrate your answer by sketching, in a single diagram, the graphs of $y = |3x - 7|$ and $y = |2x + 1|$.

(5) The function f is defined by $f(x) = 2x^2 + 2x - 7$ for $x > 0$. Show that f is 1–1 and find $f^{-1}(x)$.

Differentiation

Revision

Differentiation is an important topic in mathematics, and you will meet several new results and techniques in this module. The basic result from C1 says that

$$\text{if } y = x^n, \text{ then } \frac{dy}{dx} = nx^{n-1}$$

Check the accuracy of the following results:

$$y = 5x^2 - 7x + 9 \Rightarrow \frac{dy}{dx} = 10x - 7$$

$$y = \frac{6}{x^2} = 6x^{-2} \Rightarrow \frac{dy}{dx} = -12x^{-3}$$

$$y = \sqrt[3]{x} = x^{\frac{1}{3}} \Rightarrow \frac{dy}{dx} = \tfrac{1}{3}x^{-\frac{2}{3}}$$

Exercise 10

(1) Given that $y = 3x^4 - \dfrac{1}{x}$, find $\dfrac{dy}{dx}$.

(2) Find the gradient of the curve $y = \sqrt{x}$ at the point on the curve for which $x = 9$.

(3) Find the equation of the tangent to the curve $y = 2x^2 + 5x - 9$ at the point $(-2, -11)$.

(4) Given that $y = \dfrac{4}{x}$, find an expression for $\dfrac{d^2y}{dx^2}$.

(5) Find the stationary points of the curve $y = x^3 + 6x^2 - 36x + 11$, and determine whether each is a maximum or minimum point.

Exponential and logarithmic functions

The number e (the approximate value of which is 2.718) plays a significant part in calculus. A fundamental differentiation result for the exponential function e^x is

$$y = e^x \Rightarrow \frac{dy}{dx} = e^x$$

A more general result is

$$y = e^{ax} \Rightarrow \frac{dy}{dx} = ae^{ax}$$

where a is any constant.

Worked example 1

Given $y = 3e^{4x} + 2e^{-3x}$, find $\frac{dy}{dx}$.

Solution

$$\frac{dy}{dx} = 3 \times 4e^{4x} + 2 \times (-3)e^{-3x} = 12e^{4x} - 6e^{-3x}$$

Worked example 2

Find the equation of the tangent to the curve $y = \frac{1}{2}e^{6x}$ at the point for which $x = 0$.

Solution

Differentiating gives $\frac{dy}{dx} = 3e^{6x}$

When $x = 0$, $\frac{dy}{dx} = 3e^0 = 3$ and $y = \frac{1}{2}e^0 = \frac{1}{2}$

So the equation of the tangent is $y - \frac{1}{2} = 3(x - 0)$, i.e. $y = 3x + \frac{1}{2}$

Exercise 11

(1) Given $y = 2e^{-5x} - 4e^{\frac{1}{2}x}$, find $\frac{dy}{dx}$.

(2) Given $f(x) = 3e^{2x}$, find $f''(0)$.

(3) Find the equation of the tangent to the curve $y = 5e^{2x} + 3e^x$ at the point for which $x = 0$.

(4) Find the x-coordinate of the point on the curve $y = 2e^{-4x} + 16x - 5$ at which the gradient is zero.

The basic differentiation result for natural logarithms is

$$y = \ln x \Rightarrow \frac{dy}{dx} = \frac{1}{x}$$

This can be confirmed by using the relation $\frac{dy}{dx} = 1 \div \frac{dx}{dy}$, as follows:

The statement $y = \ln x$ is equivalent to the statement $x = e^y$

Differentiating $x = e^y$ gives $\frac{dx}{dy} = e^y$, i.e. $\frac{dx}{dy} = x$

Hence $\frac{dy}{dx} = 1 \div \frac{dx}{dy} = 1 \div x = \frac{1}{x}$

Worked example

Find $\dfrac{d}{dx}(5\ln x)$.

Solution

🔢 Note that $\dfrac{d}{dx}(5\ln x)$ is the instruction to differentiate $5\ln x$ with respect to x.

$$\frac{d}{dx}(5\ln x) = 5 \times \frac{1}{x} = \frac{5}{x}$$

Exercise 12

(1) Find $\dfrac{d}{dx}(x^3 + x - 8\ln x)$.

(2) Find the coordinates of the point on the curve $y = 3\ln x$ at which the gradient is 12.

(3) Find the x-coordinate of the stationary point on the curve $y = 2x - 6\ln x$.

Chain rule

The chain rule states that $\dfrac{dy}{dx} = \dfrac{dy}{du} \times \dfrac{du}{dx}$; this provides a technique for differentiating more complicated expressions.

Consider $y = (3x^2 + 5)^7$. The quadratic function $3x^2 + 5$ has been raised to the power 7.

Putting $u = 3x^2 + 5$ means that $y = u^7$

Differentiating both expressions in the previous line gives $\dfrac{du}{dx} = 6x$ and $\dfrac{dy}{du} = 7u^6$

Using the chain rule, we obtain $\dfrac{dy}{dx} = \dfrac{dy}{du} \times \dfrac{du}{dx} = 7u^6 \times 6x$

Replacing u by $3x^2 + 5$ gives $\dfrac{dy}{dx} = 7(3x^2 + 5)^6 \times 6x = 42x(3x^2 + 5)^6$

In practice, it is best to apply the chain rule by just keeping the 'u' in mind without the fuss of actually introducing it on paper. Observe how the chain rule operates in each of the following examples. Make sure you can identify the u in each case.

$$y = (5x + 9)^8 \Rightarrow \frac{dy}{dx} = 8(5x + 9)^7 \times 5 = 40(5x + 9)^7$$

$$y = e^{4x+3} \Rightarrow \frac{dy}{dx} = e^{4x+3} \times 4 = 4e^{4x+3}$$

$$y = \ln(3x + 1) \Rightarrow \frac{dy}{dx} = \frac{1}{3x+1} \times 3 = \frac{3}{3x+1}$$

$$y = \frac{1}{x^3 + 2} = (x^3 + 2)^{-1} \Rightarrow \frac{dy}{dx} = -1(x^3 + 2)^{-2} \times 3x^2 = -\frac{3x^2}{(x^3+2)^2}$$

$$y = \sqrt{5e^{2x} + 1} = (5e^{2x} + 1)^{\frac{1}{2}} \Rightarrow \frac{dy}{dx} = \tfrac{1}{2}(5e^{2x} + 1)^{-\frac{1}{2}} \times 10e^{2x} = \frac{5e^{2x}}{\sqrt{5e^{2x}+1}}$$

Worked example

Find the gradient of the curve $y = \frac{1}{2}(x^2 - 5)^4$ at the point for which $x = 3$.

Solution

Differentiating, $\dfrac{dy}{dx} = \frac{1}{2} \times 4(x^2 - 5)^3 \times 2x = 4x(x^2 - 5)^3$

When $x = 3$, $\dfrac{dy}{dx} = 4 \times 3 \times (9 - 5)^3 = 768$

> **n** Here, u is $x^2 - 5$; differentiating it gives $2x$. Be careful that you do not omit this part of the chain rule.

Exercise 13

(1) Given $y = (4x - 5)^{10}$, find $\dfrac{dy}{dx}$.

(2) Given $y = \ln(x^2 + 4)$, find $\dfrac{dy}{dx}$.

(3) Given $y = (e^{3x} + 2)^6$, find $\dfrac{dy}{dx}$.

(4) Given $y = \dfrac{4}{x^2 + x + 1}$, find $\dfrac{dy}{dx}$.

(5) Given $y = \sqrt{8x + 1}$, find $\dfrac{d^2y}{dx^2}$.

Product rule

The product rule provides the technique for differentiating expressions such as $x^2(2x - 1)^5$ and $2x^7 \ln x$. Note that each of these consists of two separate functions of x multiplied together. The product rule says that

if $y = uv$ where u and v are functions of x, then $\dfrac{dy}{dx} = \dfrac{du}{dx} \times v + u \times \dfrac{dv}{dx}$

For example, if $y = x^3 e^{5x}$, then $\dfrac{dy}{dx} = 3x^2 \times e^{5x} + x^3 \times 5e^{5x}$

The process is: (x^3 differentiated) × (e^{5x}) plus (x^3) × (e^{5x} differentiated).

It is preferable to remember the process rather than memorise the actual statement of the product rule. Be alert both for expressions which will require the product rule and for those where it is not needed. Can you identify the three expressions in the following list which will need the product rule for differentiation?

$6x(2x - 1)^{11}$ $\qquad$ $18(x^2 + 6)^{\frac{1}{3}}$ $\qquad$ $4e^{x-1}$

$e^{4x} \ln(1 - x)$ $\qquad$ $\ln(x^2 + 4x + 1)$ $\qquad$ $(2 - x)^3 (3 - x)^4$

> **Worked example 1**
>
> Differentiate $5x^3(2x + 1)^4$ with respect to x.
>
> **Solution**
>
> Using the product rule, the derivative is $15x^2(2x + 1)^4 + 5x^3 \times 4(2x + 1)^3 \times 2 = 15x^2(2x + 1)^4 + 40x^3(2x + 1)^3$
>
> **n** The chain rule is used when $(2x + 1)^4$ is differentiated to give $4(2x + 1)^3 \times 2$. The final step involves some straightforward simplification.

Worked example 2

Find the coordinates of the stationary point on the curve $y = 4xe^{-2x}$.

Solution

Differentiating by the product rule gives $4e^{-2x} + 4x(-2e^{-2x}) = 4e^{-2x} - 8xe^{-2x}$
$$= 4e^{-2x}(1 - 2x)$$

At a stationary point, $\frac{dy}{dx} = 0$, i.e. $4e^{-2x}(1 - 2x) = 0$

Since e^{-2x} can never be zero, the stationary point must have $1 - 2x = 0$, giving $x = \frac{1}{2}$

When $x = \frac{1}{2}, y = 4 \times \frac{1}{2} \times e^{-1} = 2e^{-1}$

So the stationary point is $(\frac{1}{2}, 2e^{-1})$

> **11** You are expected to give the exact values of coordinates unless a question states otherwise. So here the y-coordinate is given as $2e^{-1}$ rather than in decimal form 0.736, which is merely an approximation to the true value.

Exercise 14

(1) Given $y = x^3 \ln x$, find $\frac{dy}{dx}$.

(2) Given $y = x^2(3x + 1)^{\frac{1}{2}}$, find $\frac{dy}{dx}$.

(3) Find the equation of the tangent to the curve $y = x^2(2x - 1)^5$ at the point $(1, 1)$.

Quotient rule

The quotient rule says that

if $y = \frac{u}{v}$ where u and v are functions of x, then $\frac{dy}{dx} = \frac{v \cdot \frac{du}{dx} - u \cdot \frac{dv}{dx}}{v^2}$

This result provides the process for differentiating expressions such as $\frac{2x+5}{x-4}$, $\frac{4e^{5x}}{x}$ and $\frac{\ln(2x+1)}{x^2}$.

There is no need to memorise the quotient rule for the examination because it is included in the *List of Formulae*. (Note, though, that it is given there in a different form: if $y = \frac{f(x)}{g(x)}$, then $\frac{dy}{dx} = \frac{f'(x)g(x) - f(x)g'(x)}{\{g(x)\}^2}$.) Note that the quotient rule is not needed for differentiating something like $y = \frac{6}{(2x+5)^3}$, as this can be rewritten in the form $y = 6(2x + 5)^{-3}$, from which $\frac{dy}{dx} = -36(2x + 5)^{-4}$ follows immediately.

Worked example 1

Given $y = \frac{4x+1}{2x+3}$, find $\frac{dy}{dx}$.

Solution

Here $u = 4x + 1$ and hence $\dfrac{du}{dx} = 4$

Also, $v = 2x + 3$ and hence $\dfrac{dv}{dx} = 2$

The quotient rule then gives $\dfrac{dy}{dx} = \dfrac{(2x+3)\cdot 4 - (4x+1)\cdot 2}{(2x+3)^2} = \dfrac{8x+12-8x-2}{(2x+3)^2} = \dfrac{10}{(2x+3)^2}$

 Some simplification was possible here, revealing that the gradient of the $y = \dfrac{4x+1}{2x+3}$ curve is always positive. Can you see why?

Worked example 2

Find the x-coordinate of the stationary point on the curve $y = \dfrac{e^{3x}}{x+1}$.

Solution

Using the quotient rule, $\dfrac{dy}{dx} = \dfrac{(x+1)\cdot 3e^{3x} - e^{3x}\cdot 1}{(x+1)^2}$

Simplifying gives $\dfrac{dy}{dx} = \dfrac{(3x+2)e^{3x}}{(x+1)^2}$

For a stationary point, $\dfrac{dy}{dx} = 0$, so it must have $(3x + 2)e^{3x} = 0$

Since e^{3x} can never be zero, it follows that $3x + 2 = 0$, i.e. $x = -\dfrac{2}{3}$

Exercise 15

(1) Given $y = \dfrac{4x}{x^2 + 1}$, find $\dfrac{dy}{dx}$.

(2) Given $f(x) = \dfrac{\ln x}{x + 2}$, find the exact value of $f'(e)$.

(3) Find the coordinates of the stationary points on the curve with equation $y = \dfrac{x - 1}{x^2 + 3}$.

Exponential growth and decay

Consider the equation $P = 200e^{0.04t}$, where P represents the size of a population of bacteria at time t hours (with $t \geq 0$). The graph of $P = 200e^{0.04t}$ is as follows:

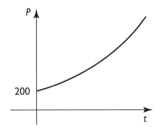

This population is said to be growing exponentially. There are two important properties of exponential growth. The first property is that **over equal stretches of time**, the population **increases by the same factor**.

t	0	12	24	36	48
P	200	323	522	844	1364

The table shows values of P at 12-hour intervals; each value of P is approximately 1.616 times the previous value.

The second property of exponential growth concerns the rate of increase of P. The rate of increase is given by the derivative $\dfrac{dP}{dt}$, which in this example is $8e^{0.04t}$. Note that $8e^{0.04t}$ is the same as $0.04 \times 200e^{0.04t}$, that is, $0.04P$. Thus, **the rate of increase of P is proportional to P itself**.

Exponential growth, with these two properties, is also exhibited by equations such as $Q = 150 \times 2^{\frac{1}{4}t}$.

Exponential decay is shown by equations such as $S = 40e^{-2t}$ and $V = 320 \times 3^{-\frac{1}{2}t}$; two analogous properties hold, with 'increase' replaced by 'decrease'.

Worked example

A substance is decaying exponentially so that its mass, m grams, after t years is given by the formula $m = 250e^{-0.07t}$.

(i) Find the number of years taken for the mass to decrease by half.

(ii) Find the rate at which the mass is decreasing after 40 years.

Solution

(i) The initial mass, when $t = 0$, is 250 grams.

When the mass is halved to 125 grams, we have $125 = 250e^{-0.07t}$

Dividing through by 250 gives $0.5 = e^{-0.07t}$

Taking natural logarithms, $\ln 0.5 = -0.07t$ and hence $t = -\dfrac{1}{0.07}\ln 0.5 \approx 9.90$

Therefore it takes 9.9 years for the mass to decrease by 50%.

(ii) $m = 250e^{-0.07t} \Rightarrow \dfrac{dm}{dt} = -17.5e^{-0.07t}$

When $t = 40$, $\dfrac{dm}{dt} = -17.5e^{-0.07 \times 40} \approx -1.06$

So, after 40 years, the mass is decreasing at a rate of 1.1 grams per year (to 2 significant figures).

It takes 9.9 years for the mass to decrease from 250 g to 125 g. The first property of exponential growth/decay implies that it will also take 9.9 years for the mass to decrease from 80 g to 40 g or from 10 g to 5 g. There is always an element of approximation in a problem like this, thus it would be inappropriate to give answers to more than 2 or 3 significant figures. Note that the calculation concludes with $\frac{dm}{dt} \approx -1.06$; the negative sign indicates that this is a rate of decrease (rather than increase).

Exercise 16

(1) The size Q of a certain population after t years is given by $Q = 700e^{0.018t}$. Find the time taken for the population to double.

(2) The mass, m kg, of a certain substance at time t days is given by $m = 48e^{-0.05t}$. Find the value of t for which the mass is decreasing at a rate of 0.8 kg per day.

(3) A population P is increasing exponentially. The table shows values of P after t years.

t	0	8	16	24	48
P	120	180			

Complete the table.

Connected rates of change

If we know the rate at which the volume of a sphere is increasing, is it then possible to determine the rate at which the radius is increasing? As an example, suppose the volume is increasing at 350 cm³ per minute, i.e. $\frac{dV}{dt} = 350$. We know that $V = \frac{4}{3}\pi r^3$, so $\frac{dV}{dr} = 4\pi r^2$. Can we find $\frac{dr}{dt}$, the rate of increase of the radius?

The chain rule says that $\frac{dV}{dt} = \frac{dV}{dr} \times \frac{dr}{dt}$, so in this case we have $350 = 4\pi r^2 \times \frac{dr}{dt}$. Hence, for a particular value of r, we can determine the corresponding value of $\frac{dr}{dt}$. For example, at the instant when $r = 12$, we have $350 = 4\pi \times 12^2 \times \frac{dr}{dt}$, giving $\frac{dr}{dt} = \frac{350}{4\pi \times 144} \approx 0.19$. So, at the instant when the radius is 12 cm, it is increasing at a rate of 0.19 cm per minute.

Worked example
Earth is being added to a mound in such a way that when the height of the mound is h metres, its mass, M kilograms, is given by $M = 420h^2 + 85h^3$. Assuming that earth is being added at a rate of 1400 kg per hour, find the rate at which the height is increasing at the instant when $h = 8$.

Solution
$M = 420h^2 + 85h^3 \Rightarrow \frac{dM}{dh} = 840h + 255h^2$

When $h = 8$, $\frac{dM}{dh} = 840 \times 8 + 255 \times 8^2 = 23040$

The question tells us that the rate of increase of the mass is 1400 kg per hour, i.e. $\frac{dM}{dt} = 1400$, where t is the time in hours.

We need to find $\dfrac{dh}{dt}$

Using the chain rule, $\dfrac{dM}{dt} = \dfrac{dM}{dh} \times \dfrac{dh}{dt}$, so $1400 = 23040 \dfrac{dh}{dt}$ and hence $\dfrac{dh}{dt} = \dfrac{1400}{23040}$

≈ 0.0608

Thus, the height is increasing at 0.061 metres per hour.

Exercise 17

(1) Air is being released from a spherical balloon at a rate of $450\,\text{cm}^3\,\text{s}^{-1}$. Find the rate at which the radius is decreasing at the instant when the radius is 60 cm.

(2) Oil is leaking and forming a circular patch, which is expanding in such a way that the radius increases by 0.8 metres per day. Find the rate at which the area of the patch is increasing at the instant when the radius is 20 metres.

General differentiation problems

Exercise 18

(1) Find the coordinates of the point on the curve $y = (4x + 3)^{\frac{3}{2}}$ at which the gradient is $30\sqrt{3}$.

(2) Find the x-coordinates of the stationary points on the curve $y = 4x^2 e^{-3x}$.

(3) Find the equation of the normal to the curve $y = \dfrac{5x}{x^2 + 1}$ at the point $(3, \frac{3}{2})$.

(4) A quantity Q is increasing exponentially. At the start its value is 120, and every 12 days its value doubles.

    (i) Find the value of Q 60 days after the start.

    (ii) Find a formula for Q in terms of t, where t is the number of days after the start.

    (iii) Find the number of days after the start at which the value of Q is 85000.

Integration

Revision

You were introduced to integration in C2, and the topic is further developed in this module. The basic result from C2 is

$$\int x^n\,dx = \frac{1}{n+1} x^{n+1} + c$$

Use of this formula gives the following:

$$\int (2x^2 - 6x + 4)\,dx = \tfrac{2}{3}x^3 - 3x^2 + 4x + c$$

$$\int_4^9 \sqrt{x}\,dx = \int_4^9 x^{\frac{1}{2}}\,dx = \left[\tfrac{2}{3}x^{\frac{3}{2}}\right]_4^9 = \tfrac{2}{3} \times 9^{\frac{3}{2}} - \tfrac{2}{3} \times 4^{\frac{3}{2}} = \tfrac{2}{3} \times 27 - \tfrac{2}{3} \times 8 = \tfrac{38}{3}$$

Exercise 19

(1) Find $\int (8x^3 + 1)\,dx$.

(2) Find $\int (\dfrac{3}{\sqrt{x}} - x)\,dx$.

(3) Evaluate $\int_1^3 (2x+1)(2x+3)\,dx$.

(4) Find the area of the region enclosed by the curve $y = \dfrac{10}{x^2}$ and the lines $x = 1, x = 2$ and $y = 0$.

Integration of $(ax + b)^n$

As integration is the reverse process of differentiation, each differentiation result leads to a corresponding integration result.

The chain rule for differentiation gives $y = \dfrac{1}{a(n+1)}(ax+b)^{n+1} \rightarrow \dfrac{dy}{dx} = (ax+b)^n$

It follows that

$$\int (ax+b)^n \, dx = \frac{1}{a(n+1)}(ax+b)^{n+1} + c$$

Note that upon integrating the power increases by one; also, you must divide by the new power and by the coefficient a.

Confirm the following results:

$$\int (2x+5)^7 \, dx = \tfrac{1}{16}(2x+5)^8 + c$$

$$\int \sqrt{4x-1} \, dx = \int (4x-1)^{\frac{1}{2}} \, dx = \frac{1}{4 \times \frac{3}{2}}(4x-1)^{\frac{3}{2}} + c = \tfrac{1}{6}(4x-1)^{\frac{3}{2}} + c$$

$$\int \frac{8}{(5x+1)^3} \, dx = \int 8(5x+1)^{-3} \, dx = \frac{8}{5 \times (-2)}(5x+1)^{-2} + c = -\tfrac{4}{5}(5x+1)^{-2} + c$$

Worked example

Find the value of $\int_0^{13} \dfrac{6}{\sqrt[3]{2x+1}} \, dx$.

Solution

$$\int_0^{13} \frac{6}{\sqrt[3]{2x+1}}\,dx = \int_0^{13} 6(2x+1)^{-\frac{1}{3}}\,dx = \left[\frac{6}{2 \times \frac{2}{3}}(2x+1)^{\frac{2}{3}}\right]_0^{13} = \left[\tfrac{9}{2}(2x+1)^{\frac{2}{3}}\right]_0^{13} = \tfrac{9}{2} \times 27^{\frac{2}{3}} - \tfrac{9}{2} \times 1^{\frac{2}{3}}$$

$$= \tfrac{81}{2} - \tfrac{9}{2} = 36$$

n You must be sure of results about indices from C1. Do not make the mistake of jumping to the conclusion that because the lower limit of the integration is zero, the value obtained by substituting that limit will automatically be zero.

Exercise 20

(1) Find $\int 10(2x-3)^9 \, dx$.

(2) Find $\int \dfrac{1}{\sqrt{8x+3}} \, dx$.

(3) Find the exact value of $\int_1^2 \dfrac{4}{(2x+1)^2} \, dx$.

Integration of $(ax + b)^{-1}$

The formula $\int x^n \, dx = \dfrac{1}{n+1} x^{n+1} + c$ works for all values of n except $n = -1$. Now we can deal with that special case. Since the derivative of $\ln x$ is $\dfrac{1}{x}$, we have the results

$$\int \frac{1}{x} \, dx = \ln|x| + c$$

$$\int \frac{1}{ax+b} \, dx = \frac{1}{a} \ln|ax+b| + c$$

The modulus signs are needed so that the logarithm is well defined when negative values of x or $ax + b$ are involved.

Worked example 1

Find the exact value of $\int_1^4 \dfrac{3+x}{x} \, dx$.

Solution

$$\int_1^4 \frac{3+x}{x} \, dx = \int_1^4 \left(\frac{3}{x} + 1 \right) dx = \left[3\ln|x| + x \right]_1^4 = 3\ln 4 + 4 - (3\ln 1 + 1) = 3\ln 4 + 3$$

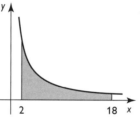 Note the first step, rewriting $\dfrac{3+x}{x}$ as $\dfrac{3}{x} + 1$, which is necessary before integration can be attempted.

Worked example 2

Find the exact value of $\int_{-7}^{-1} \dfrac{1}{2x-1} \, dx$.

Solution

$$\int_{-7}^{-1} \frac{1}{2x-1} \, dx = \left[\tfrac{1}{2}\ln|2x-1| \right]_{-7}^{-1} = \tfrac{1}{2}\ln|-3| - \tfrac{1}{2}\ln|-15| = \tfrac{1}{2}\ln 3 - \tfrac{1}{2}\ln 15 = \tfrac{1}{2}\ln\tfrac{3}{15} = \tfrac{1}{2}\ln\tfrac{1}{5}$$

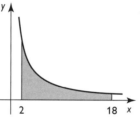 The modulus signs are needed for the evaluation of $\ln|-3|$ and $\ln|-15|$. The answer could also be given as $-\tfrac{1}{2}\ln 5$; can you see why?

Exercise 21

(1) Find the value of $\int_0^4 \dfrac{9}{3x+2} \, dx$, giving your answer in the form $\ln p$.

(2) The diagram shows part of the curve $y = \dfrac{1}{2x}$. Find the area of the shaded region.

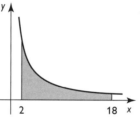

Exponential functions

The basic result is

$$\int e^{ax}\, dx = \frac{1}{a} e^{ax} + c$$

Differentiation of the right-hand side easily confirms that this formula is correct.

Worked example 1

Find $\int (6e^{2x} - e^{-x})\, dx$.

Solution

Applying the basic result,

$$\int (6e^{2x} - e^{-x})\, dx = 6 \times \tfrac{1}{2} e^{2x} - \tfrac{1}{(-1)} e^{-x} + c = 3e^{2x} + e^{-x} + c$$

 As with any integration result, the answer can be easily checked. If the result is differentiated, the original $6e^{2x} - e^{-x}$ should be obtained. Also, for an indefinite integral such as this, do not forget to include the arbitrary constant c.

Worked example 2

Find the exact value of $\int_{2}^{4} 3e^{\frac{1}{2}x-1}\, dx$.

Solution

$$\int_{2}^{4} 3e^{\frac{1}{2}x-1}\, dx = \left[3 \times \frac{1}{\frac{1}{2}} e^{\frac{1}{2}x-1} \right]_{2}^{4} = \left[6e^{\frac{1}{2}x-1} \right]_{2}^{4} = 6e^{1} - 6e^{0} = 6e - 6$$

 The request for the exact value means that we must leave the answer in terms of e.

Exercise 22

(1) Find $\int (6e^{3x} + 4e^{-2x})\, dx$.

(2) Evaluate $\int_{-1}^{0} 10e^{-2x}\, dx$.

Volumes of revolution

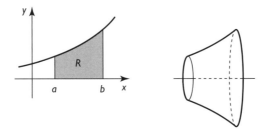

If the shaded region R is rotated completely about the x-axis, a solid is produced. The volume of this solid is given by

$$V = \int_a^b \pi\, y^2\, dx$$

where y must be expressed in terms of x.

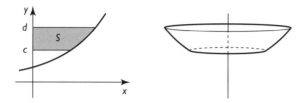

If the shaded region S is rotated completely about the y-axis, a solid is produced. The volume of this solid is given by

$$V = \int_c^d \pi\, x^2\, dy$$

where x must be expressed in terms of y.

Worked example 1

The region enclosed by the curve $y = x^3$ and the lines $x = 1$, $x = 2$ and $y = 0$ is rotated completely about the x-axis. Find the volume of the solid produced.

Solution

Volume $= \int_1^2 \pi\, y^2\, dx = \int_1^2 \pi(x^3)^2\, dx = \int_1^2 \pi x^6\, dx = \left[\frac{1}{7}\pi x^7\right]_1^2 = \frac{128}{7}\pi - \frac{1}{7}\pi = \frac{127}{7}\pi$

Worked example 2

The region R is enclosed by the curve $y = 2e^{-x}$ and the lines $x = 0$, $x = 1$ and $y = 0$. Find the exact volume of the solid produced when R is rotated completely about the x-axis.

Solution

11 While drawing a graph is not essential, it does help you to see what is involved.

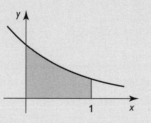

Volume $= \int_0^1 \pi(2e^{-x})^2\, dx = \int_0^1 4\pi e^{-2x}\, dx = \left[-2\pi e^{-2x}\right]_0^1 = -2\pi e^{-2} - (-2\pi) = 2\pi(1 - e^{-2})$

It is important to find y^2 in terms of x carefully before attempting integration. The question asks for the exact volume, which means that the answer must be left in terms of π and e.

Exercise 23

(1) The region bounded by the curve $y = 3\sqrt{x}$ and the lines $x = 2$ and $y = 0$ is rotated completely about the x-axis. Find the volume of the solid produced.

(2)

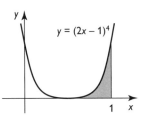

The shaded region is rotated completely about the x-axis. Find the volume of the solid generated.

(3)

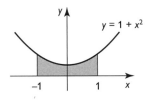

Find the exact volume of the solid produced when the shaded region is rotated completely about the line $y = 0$.

Simpson's rule

In module C2 you met the trapezium rule, a numerical approach to integration. In C3, Simpson's rule is covered, which provides a means of finding an approximation to the value of an integral. This is especially useful in cases where the integral cannot be found in the usual way. The formula for Simpson's rule is given in the *List of Formulae*:

$$\int_a^b y \, dx \approx \tfrac{1}{3}h\{(y_0 + y_n) + 4(y_1 + y_3 + \ldots + y_{n-1}) + 2(y_2 + y_4 + \ldots + y_{n-2})\},$$

where $h = \dfrac{b-a}{n}$ and n is even.

The number of strips, represented by n, must be an even number. The width of each strip is h.

If there are two strips, then $\int_a^b y \, dx \approx \tfrac{1}{3}h(y_0 + 4y_1 + y_2)$

If there are four strips, then $\int_a^b y \, dx \approx \tfrac{1}{3}h(y_0 + 4y_1 + 2y_2 + 4y_3 + y_4)$

If there are six strips, then $\int_a^b y \, dx \approx \tfrac{1}{3}h(y_0 + 4y_1 + 2y_2 + 4y_3 + 2y_4 + 4y_5 + y_6)$

Worked example

Use Simpson's rule with four strips to find an approximation to $\int_0^8 \ln(x^2+5)\,dx$.

Solution

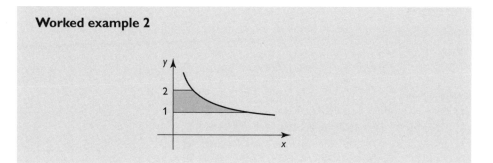 $\ln(x^2 + 5)$ is such that a formula for its indefinite integral cannot be found, i.e. there is no algebraic expression which can be differentiated to give $\ln(x^2 + 5)$. Simpson's rule will provide a very good approximation to the definite integral.

Here $n = 4$ and $h = \frac{8-0}{4} = 2$, so we need to use values of $\ln(x^2 + 5)$ that correspond to the x-values 0, 2, 4, 6 and 8, namely $\ln 5$, $\ln 9$, $\ln 21$, $\ln 41$ and $\ln 69$

Using Simpson's rule, $\int_0^8 \ln(x^2+5)\,dx \approx \frac{1}{3} \times 2(\ln 5 + 4\ln 9 + 2\ln 21 + 4\ln 41 + \ln 69)$

giving $\int_0^8 \ln(x^2+5)\,dx \approx 23.7$ correct to 3 significant figures.

Note the pattern of the coefficients for the y-values: 1, 4, 2, 4, 1.

Exercise 24

(1) Use Simpson's rule with six strips to find an approximation to $\int_0^6 \sqrt{2+x^2}\,dx$.

(2) Use Simpson's rule with two strips to find an approximation to $\int_1^5 \frac{10}{x^3+3}\,dx$.

General integration problems

Worked example 1

Find the exact value of $\int_1^\infty 4e^{-2x}\,dx$.

Solution

$$\int_1^\infty 4e^{-2x}\,dx = \left[-2e^{-2x}\right]_1^\infty$$

The upper limit is 'infinity'. For large positive values of x, $-2e^{-2x}$, which can be written as $-\dfrac{2}{e^{2x}}$, is very close to zero. So, in the limit as x goes to ∞, $-2e^{-2x}$ becomes zero.

Hence $\left[-2e^{-2x}\right]_1^\infty = 0 - (-2e^{-2}) = 2e^{-2}$

Worked example 2

The diagram shows part of the curve $y = \dfrac{2}{\sqrt{x}}$. The shaded region is rotated completely about the y-axis. Find the volume of the solid produced.

Solution

The volume is $\int_1^2 \pi x^2 \, dy$, where x must be expressed in terms of y

From $y = \dfrac{2}{\sqrt{x}}$ we obtain $\sqrt{x} = \dfrac{2}{y}$ and thus $x^2 = \dfrac{16}{y^4}$

So volume $= \displaystyle\int_1^2 \pi \cdot \dfrac{16}{y^4} \, dy = \int_1^2 16\pi \, y^{-4} dy = \left[\dfrac{16\pi \, y^{-3}}{-3}\right]_1^2 = -\tfrac{16}{24}\pi - (-\tfrac{16}{3}\pi) = \tfrac{14}{3}\pi$

Exercise 25

(1) Given that $\displaystyle\int_1^a \dfrac{3}{2x+7} \, dx = \ln 27$, find the value of a.

(2) Find the exact value of $\displaystyle\int_0^\infty \dfrac{4+e^x}{e^{3x}} \, dx$.

(3) Use Simpson's rule with four strips to find an approximation to $\displaystyle\int_0^{20} \tan^{-1} x \, dx$.

(4) The region shaded in the diagram is rotated completely about the y-axis. Find the exact volume of the solid generated.

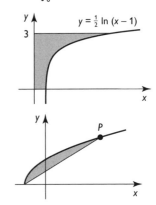

$y = \tfrac{1}{2} \ln (x-1)$

(5) The diagram shows the curve $y = \sqrt{3x+7}$. At the point P, the gradient of the curve is 0.3. Find the exact area of the shaded region.

Trigonometry

Revision

You became acquainted with some of the properties of sine, cosine and tangent in C2. Two important identities are:

$$\sin^2 \theta + \cos^2 \theta \equiv 1 \qquad \tan \theta \equiv \dfrac{\sin \theta}{\cos \theta}$$

Exercise 26

(1) Express $3\cos^2 \theta + 4\sin \theta \cos \theta \tan \theta$ in terms of $\sin \theta$.

(2) Solve, for $0° < \theta < 360°$, the equation $4 \sin \theta = 9 \cos \theta$.

(3) Solve, for $0° < \theta < 360°$, the equation $6 \sin^2 \theta + \cos \theta = 4$.

Reciprocal trigonometric functions

Secant, cosecant and cotangent are three more trigonometric functions. They are defined as follows:

$$\sec \theta = \frac{1}{\cos \theta} \qquad\qquad \operatorname{cosec} \theta = \frac{1}{\sin \theta} \qquad\qquad \cot \theta = \frac{1}{\tan \theta}$$

Note how the graphs of each pair of trigonometric functions are related. Each graph is shown for $-360° < \theta < 360°$.

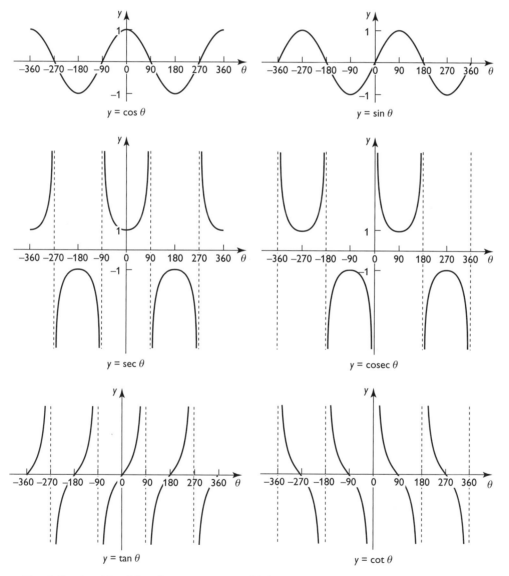

The following identities (i.e. statements which are true for all possible values of the angle) show how different trigonometric functions are linked:

$$\cot\theta \equiv \frac{\cos\theta}{\sin\theta} \qquad \sec^2\theta \equiv 1 + \tan^2\theta \qquad \csc^2\theta \equiv 1 + \cot^2\theta$$

These identities are useful for solving certain trigonometric equations.

Worked example 1

Solve, for $0° < \theta < 180°$, the equation $\sec 2\theta = 5$.

Solution

The equation $\sec 2\theta = 5$ is equivalent to $\dfrac{1}{\cos 2\theta} = 5$, which leads to $\cos 2\theta = \frac{1}{5}$

Hence $2\theta = 78.46°,\ 281.54°$ (with $0° < 2\theta < 360°$)

and so $\theta = 39.2°,\ 140.8°$

$\boxed{n}$ To ensure that we collect all the answers between $0°$ and $180°$ as requested, we must first find values of 2θ that are between $0°$ and $360°$.

Worked example 2

Solve, for $0 < x < \pi$, the equation $\sec^2 x = 3\tan x + 5$.

Solution

$\boxed{n}$ The identity $\sec^2 x \equiv 1 + \tan^2 x$ can be used to convert the given equation into one in which only $\tan x$ occurs. Notice that answers in radians are requested, so your calculator needs to be in radian mode.

Using the identity, $\sec^2 x = 3\tan x + 5$ becomes $1 + \tan^2 x = 3\tan x + 5$

Simplifying, $\tan^2 x - 3\tan x - 4 = 0$

Factorising, $(\tan x - 4)(\tan x + 1) = 0$

Hence $\tan x = 4$ or $\tan x = -1$, giving $x = 1.33$ or $x = \frac{3}{4}\pi$

$\boxed{n}$ The equation $\tan x = 4$ gives $x = 1.33$ (to 3 significant figures), which is a value in the first quadrant; $\tan x = -1$ gives the exact solution $x = \frac{3}{4}\pi$, a value in the second quadrant.

Exercise 27

(1) Solve, for $0° < \theta < 360°$, the equation $\csc\frac{1}{2}\theta = 6$.

(2) Solve, for $0° < \theta < 360°$, the equation $2\cot\theta = 3\cos\theta$.

(3) Solve, for $0° < \theta < 360°$, the equation $\tan^2\theta + 9 = 6\sec\theta$.

Further trigonometric identities

Six useful identities — sometimes called the angle sum and difference identities — are the following:

$$\sin(A + B) \equiv \sin A \cos B + \cos A \sin B \qquad \sin(A - B) \equiv \sin A \cos B - \cos A \sin B$$

$$\cos(A + B) \equiv \cos A \cos B - \sin A \sin B \qquad \cos(A - B) \equiv \cos A \cos B + \sin A \sin B$$

$$\tan(A + B) \equiv \frac{\tan A + \tan B}{1 - \tan A \tan B} \qquad \tan(A - B) \equiv \frac{\tan A - \tan B}{1 + \tan A \tan B}$$

You do not have to memorise these identities because they are included in the *List of Formulae*. There, each of the three pairs is expressed in one formula; for example, the final two identities are put together as $\tan(A \pm B) \equiv \dfrac{\tan A \pm \tan B}{1 \mp \tan A \tan B}$. Note how, in each combined formula, the + and − signs must be selected consistently from the ± and ∓ to pick out each of the two identities.

Worked example 1

Given that A and B are acute angles such that $\sin A = \frac{1}{3}$ and $\sin B = \frac{2}{3}$, find the exact value of $\cos(A + B)$.

Solution

🔢 To use the identity for $\cos(A + B)$, the values of $\cos A$ and $\cos B$ are needed. The important identity $\sin^2 \theta + \cos^2 \theta \equiv 1$ will help.

$\sin^2 A + \cos^2 A = 1$, giving $\frac{1}{9} + \cos^2 A = 1$ and hence $\cos^2 A = \frac{8}{9}$
Since A is acute, $\cos A$ is positive and so $\cos A = \frac{1}{3}\sqrt{8}$
Similarly, $\sin^2 B + \cos^2 B = 1$ gives $\frac{4}{9} + \cos^2 B = 1$ and hence $\cos B = \frac{1}{3}\sqrt{5}$
Substituting these values into $\cos(A + B) = \cos A \cos B - \sin A \sin B$ yields
$\cos(A + B) = \frac{1}{3}\sqrt{8} \times \frac{1}{3}\sqrt{5} - \frac{1}{3} \times \frac{2}{3} = \frac{1}{9}(\sqrt{40} - 2) = \frac{2}{9}(\sqrt{10} - 1)$

Worked example 2

Given that $\tan(\theta + 45°) = 6$, find the value of $\tan \theta$.

Solution

Using the identity for $\tan(A + B)$, the given equation $\tan(\theta + 45°) = 6$ becomes
$\dfrac{\tan \theta + \tan 45°}{1 - \tan \theta \tan 45°} = 6$
Remembering that $\tan 45° = 1$, we have $\dfrac{\tan \theta + 1}{1 - \tan \theta} = 6$
Multiplying both sides by $1 - \tan \theta$, we get $\tan \theta + 1 = 6(1 - \tan \theta)$
which gives $\tan \theta + 1 = 6 - 6\tan \theta$ and hence $\tan \theta = \frac{5}{7}$

🔢 Here the result $\tan 45° = 1$ was needed. It is useful to remember the sine, cosine and tangent values for the particular angles 30°, 45° and 60° — such as $\sin 45° = \dfrac{1}{\sqrt{2}}$ and $\cos 30° = \frac{1}{2}\sqrt{3}$.

Exercise 28

(1) Given that the value of $\tan \theta$ is $4\sqrt{3}$, find the exact value of $\tan(\theta - 60°)$.

(2) Show that $\cos(\theta + 60°)\cos(\theta - 60°) \equiv \cos^2 \theta - \frac{3}{4}$.

(3) It is given that θ is an acute angle such that $\cos \theta = \frac{1}{3}\sqrt{6}$. Find the exact value of $\sin(\theta + 60°)$.

Expressions of the form $a \sin \theta + b \cos \theta$

Expressions like $8\sin \theta + 6\cos \theta$, $4\sin \theta - 2\cos \theta$ or $7\cos \theta - 5\sin \theta$ crop up frequently. The identities in the previous section offer a way of writing such expressions in alternative forms. Let us take $8\sin\theta + 6\cos\theta$, for example. Consider the expansion of $R\sin(\theta + \alpha)$, where R is a positive constant:

$$R\sin(\theta + \alpha) = R\,\mathbf{sin}\,\theta\cos\alpha + R\,\mathbf{cos}\,\theta\sin\alpha$$

Compare this with $8\,\mathbf{sin}\,\theta + 6\,\mathbf{cos}\,\theta$.

The two expressions will be identical provided that we can find values of R and α for which

$$R\cos\alpha = 8 \qquad \text{and} \qquad R\sin\alpha = 6$$

Squaring and adding these two equations gives $R^2\cos^2\alpha + R^2\sin^2\alpha = 8^2 + 6^2$, i.e. $R^2(\cos^2\alpha + \sin^2\alpha) = 100$. So $R^2 \times 1 = 100$ and hence, choosing the positive root, $R = 10$.

On the other hand, dividing the two equations gives $\dfrac{R\sin\alpha}{R\cos\alpha} = \dfrac{6}{8}$, i.e. $\tan\alpha = \frac{3}{4}$, and therefore $\alpha = 36.87°$.

Hence $8\sin\theta + 6\cos\theta$ may be written in the alternative form $10\sin(\theta + 36.87°)$.

The same expression $6\cos\theta + 8\sin\theta$ can also be written in the form $R\cos(\theta - \alpha)$. Check that, in this case, $R = 10$ and $\alpha = 53.13°$.

Worked example 1

Express $\sqrt{3}\cos\theta - \sin\theta$ in the form $R\cos(\theta + \alpha)$, where $R > 0$ and $0° < \alpha < 90°$. Hence state the pair of transformations needed to transform the curve $y = \cos\theta$ to the curve $y = \sqrt{3}\cos\theta - \sin\theta$.

Solution

The expansion of $R\cos(\theta + \alpha)$ is $R\,\mathbf{cos}\,\theta\cos\alpha - R\,\mathbf{sin}\,\theta\sin\alpha$
Comparing this with $\sqrt{3}\,\mathbf{cos}\,\theta - \mathbf{sin}\,\theta$, we need to find R and α so that
$R\cos\alpha = \sqrt{3}$ and $R\sin\alpha = 1$
Squaring and adding gives $R^2 = 3 + 1$ and hence $R = 2$

Dividing gives $\tan\alpha = \dfrac{1}{\tan}$ and hence $\alpha = 30°$

Thus $\sqrt{3}\cos\theta - \sin\theta \equiv 2\cos(\theta + 30°)$
To transform $y = \cos\theta$ to $y = 2\cos(\theta + 30°)$ requires a translation by $30°$ in the negative θ-direction and a stretch in the y-direction with scale factor 2

Worked example 2

Express $3\sin\theta - 2\cos\theta$ in the form $R\sin(\theta - \alpha)$, where $R > 0$ and $0° < \alpha < 90°$. Hence solve, for $0° < \theta < 360°$, the equation $3\sin\theta - 2\cos\theta = \sqrt{10}$.

Solution

🔢 An examination question will usually indicate which of $R\sin(\theta \pm \alpha)$ and $R\cos(\theta \pm \alpha)$ is to be used in a particular case, although you should easily be able to select one with the correct expansion to match the expression under consideration.

Expanding $R\sin(\theta-\alpha)$ gives $R\sin(\theta-\alpha) = R\sin\theta\cos\alpha - R\cos\theta\sin\alpha$
Comparing with $3\sin\theta - 2\cos\theta$ means that we need $R\cos\alpha = 3$ and $R\sin\alpha = 2$
Squaring and adding leads to $R = \sqrt{13}$
Dividing gives $\tan\alpha = \frac{2}{3}$ and hence $\alpha = 33.69°$
Thus $3\sin\theta - 2\cos\theta \equiv \sqrt{13}\sin(\theta - 33.69°)$
The equation $3\sin\theta - 2\cos\theta = \sqrt{10}$ can now be rewritten as $\sqrt{13}\sin(\theta - 33.69°) = \sqrt{10}$

Dividing by $\sqrt{13}$ gives $\sin(\theta - 33.69°) = \dfrac{\sqrt{10}}{\sqrt{13}}$

Hence $\theta - 33.69° = 61.29°, 118.71°, 421.29°,...$
So $\theta = 95.0°, 152.4°$

🄝 It is important to observe how the second answer is obtained. The step

following $\sin(\theta - 33.69°) = \dfrac{\sqrt{10}}{\sqrt{13}}$ is crucial; this is the stage at which all the

potential values of $\theta - 33.69°$ must be collected. Here we have erred on the
safe side by listing one value that was not eventually needed. It is acceptable
to give angles in degrees correct to the nearest $0.1°$, unless the question
specifies otherwise.

Exercise 29

(1) Express $8\cos\theta + 15\sin\theta$ in the form $R\cos(\theta - \alpha)$, where $R > 0$ and $0° < \alpha < 90°$.
(2) Express $\sin\theta + \cos\theta$ in the form $R\sin(\theta + \alpha)$, where $R > 0$ and $0° < \alpha < 90°$, and hence sketch
the graph of $y = \sin\theta + \cos\theta$ for $-360° \le \theta \le 360°$.
(3) Express $3\cos\theta - 6\sin\theta$ in the form $R\cos(\theta + \alpha)$, where $R > 0$ and $0° < \alpha < 90°$. Hence solve,
for $0° < \theta < 360°$, the equation $3\cos\theta - 6\sin\theta = 2$.

Double-angle identities

Three important identities involving the angle 2θ are

$$\sin 2\theta \equiv 2\sin\theta\cos\theta \qquad \cos 2\theta \equiv \cos^2\theta - \sin^2\theta \qquad \tan 2\theta \equiv \frac{2\tan\theta}{1 - \tan^2\theta}$$

These do not appear in the *List of Formulae*, but they can be derived easily from the
earlier angle-sum identities applied to $\sin(\theta + \theta)$, $\cos(\theta + \theta)$ and $\tan(\theta + \theta)$, respectively.
Also note that use of the identity $\sin^2\theta + \cos^2\theta \equiv 1$ enables us to produce three versions of the identity for $\cos 2\theta$, namely

$$\cos 2\theta \equiv 2\cos^2\theta - 1 \equiv \cos^2\theta - \sin^2\theta \equiv 1 - 2\sin^2\theta$$

The double-angle identities may be adapted provided that the appropriate $2 : 1$ ratio is
maintained. For example, the following are valid identities too:

$$\sin 8\theta \equiv 2\sin 4\theta\cos 4\theta \qquad \cos\theta \equiv \cos^2\tfrac{1}{2}\theta - \sin^2\tfrac{1}{2}\theta \qquad \tan 18x \equiv \frac{2\tan 9x}{1 - \tan^2 9x}$$

Similarly,

$$\sin 40° = 2\sin 20°\cos 20° \qquad \cos 100° = 1 - 2\sin^2 50° \qquad \tan 45° = \frac{2\tan 22.5°}{1 - \tan^2 22.5°}$$

are correct statements as well.

Worked example 1

Given that θ is the acute angle such that $\cos\theta = \frac{1}{4}$, find the exact value of $\sin 2\theta$.

Solution

Using $\sin^2\theta + \cos^2\theta = 1$ and the given value of $\cos\theta$, we get $\sin^2\theta + \frac{1}{16} = 1$

Taking the positive value of $\sin\theta$ (because θ is acute) gives $\sin\theta = \frac{1}{4}\sqrt{15}$

Hence $\sin 2\theta = 2\sin\theta\cos\theta = 2 \times \frac{1}{4}\sqrt{15} \times \frac{1}{4} = \frac{1}{8}\sqrt{15}$

Worked example 2

Solve, for $0° \le \theta \le 360°$, the equation $5\sin 2\theta = 3\sin\theta$.

Solution

Using the identity for $\sin 2\theta$, the equation becomes $5 \times 2\sin\theta\cos\theta = 3\sin\theta$

i.e. $10\sin\theta\cos\theta - 3\sin\theta = 0$ and, upon factorising, $\sin\theta(10\cos\theta - 3) = 0$

Hence $\sin\theta = 0$ or $\cos\theta = \frac{3}{10}$, giving $\theta = 0°, 72.5°, 180°, 287.5°, 360°$

!! Having reached the stage $10\sin\theta\cos\theta = 3\sin\theta$, we must be careful not to simply 'cancel' $\sin\theta$, because, if we do, we will miss some of the valid answers.

Worked example 3

Given that θ is such that $3\cos 2\theta = 11\cos\theta + 7$, find the value of $\cos\theta$.

Solution

!! Of the three versions of the identity for $\cos 2\theta$, the appropriate one to use here is the one giving $\cos 2\theta$ in terms of $\cos\theta$, because then we will have an equation which involves only $\cos\theta$.

The equation becomes $3(2\cos^2\theta - 1) = 11\cos\theta + 7$

Rearranging, $6\cos^2\theta - 11\cos\theta - 10 = 0$

Factorising, $(3\cos\theta + 2)(2\cos\theta - 5) = 0$

Hence $\cos\theta = -\frac{2}{3}$

!! The factorised equation also gives $\cos\theta = \frac{5}{2}$; why is this not an acceptable answer?

Exercise 30

(1) Given that θ is an angle such that $\sin\theta = \frac{5}{6}$, find the value of $\cos 2\theta$.

(2) Solve $5\tan x \tan 2x = 8$, giving all values of x such that $0 < x < 2\pi$.

(3) Solve, for $0° \le \theta \le 360°$, the equation $4\sin 2\theta = 7\cos\theta$.

Inverse trigonometric functions

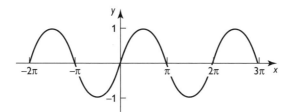

The diagram above shows the $y = \sin x$ curve. This is not the graph of a 1–1 function and so $\sin x$, for $x \in \mathbb{R}$, has no inverse. However, if the domain is restricted to $-\frac{1}{2}\pi < x < \frac{1}{2}\pi$, then the graph is as follows:

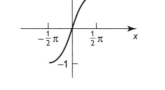

This is now 1–1 and therefore has an inverse function. The inverse function can be denoted by the equation $y = \sin^{-1} x$, and its graph is the following:

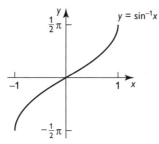

$\sin^{-1} x$ refers to the **principal value**, i.e. the value such that $-\frac{1}{2}\pi < \sin^{-1} x < \frac{1}{2}\pi$. This is the value that your calculator (set to radian mode) gives. Check that

$$\sin^{-1} 0.5 = \frac{1}{6}\pi \qquad \sin^{-1} 1 = \frac{1}{2}\pi \qquad \sin^{-1}(-0.3) = -0.305$$

The principal values of $\cos^{-1} x$ and $\tan^{-1} x$ are such that $0 \le \cos^{-1} x \le \pi$ and $-\frac{1}{2}\pi < \tan^{-1} x < \frac{1}{2}\pi$, respectively.

The corresponding graphs are shown below:

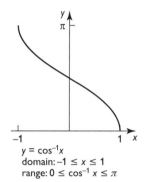

$y = \cos^{-1}x$
domain: $-1 \le x \le 1$
range: $0 \le \cos^{-1}x \le \pi$

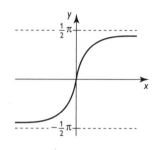

$y = \tan^{-1}x$
domain: all real numbers
range: $-\frac{1}{2}\pi < \tan^{-1}x < \frac{1}{2}\pi$

Worked example

State the value of $\tan^{-1}\sqrt{3}$.

Solution

Since $\tan\frac{1}{3}\pi = \sqrt{3}$ and $\frac{1}{3}\pi$ lies between $-\frac{1}{2}\pi$ and $\frac{1}{2}\pi$, it follows that $\tan^{-1}\sqrt{3} = \frac{1}{3}\pi$

Exercise 31

(1) State the value of $\tan^{-1}(\frac{1}{3}\sqrt{3})$.

(2) State the value of $\sin^{-1}(-\frac{1}{2}\sqrt{2})$.

(3) State the value of $\cos^{-1}(-\frac{1}{2})$.

General trigonometry problems

Worked example 1

Solve, for $0° < \theta < 360°$, the equation $4\sin(\theta - 25°) = 3\cos(\theta + 25°)$.

Solution

 It is not immediately clear how to proceed, but use of the identities for $\sin(A - B)$ and $\cos(A + B)$ would seem a sensible start.

Using the angle sum and difference identities, the given equation can be rewritten as $4(\sin\theta\cos 25° - \cos\theta\sin 25°) = 3(\cos\theta\cos 25° - \sin\theta\sin 25°)$

i.e. $4\sin\theta\cos 25° - 4\cos\theta\sin 25° = 3\cos\theta\cos 25° - 3\sin\theta\sin 25°$

Each term has either $\sin\theta$ or $\cos\theta$ involved.

Dividing through by $\cos\theta$, we have $4\tan\theta\cos 25° - 4\sin 25° = 3\cos 25° - 3\tan\theta\sin 25°$

Rearranging gives $4\tan\theta\cos 25° + 3\tan\theta\sin 25° = 3\cos 25° + 4\sin 25°$

Taking $\tan\theta$ as the common factor in the left-hand side,

$\tan\theta(4\cos 25° + 3\sin 25°) = 3\cos 25° + 4\sin 25°$

Hence $\tan\theta = \dfrac{3\cos 25° + 4\sin 25°}{4\cos 25° + 3\sin 25°}$

i.e. $\tan\theta = 0.901148$, giving $\theta = 42.0°, 222.0°$

Worked example 2

Use the identity for $\tan 2\theta$ to find the exact value of $\tan 22\frac{1}{2}°$.

Solution

The identity to use is $\tan 2\theta \equiv \dfrac{2\tan\theta}{1-\tan^2\theta}$

Putting $\theta = 22\frac{1}{2}°$, we obtain $\tan 45° = \dfrac{2\tan 22\frac{1}{2}°}{1-\tan^2 22\frac{1}{2}°}$

Remembering that $\tan 45° = 1$ and writing $\tan 22\frac{1}{2}° = t$, we have $1 = \dfrac{2t}{1-t^2}$,

leading to $t^2 + 2t - 1 = 0$

Solving this quadratic equation by using the formula, we get

$$t = \frac{-2\pm\sqrt{4+4}}{2} = \frac{-2\pm\sqrt{8}}{2} = \frac{-2\pm 2\sqrt{2}}{2} = -1\pm\sqrt{2}$$

Since $\tan 22\frac{1}{2}°$ is obviously positive, we choose $t = -1+\sqrt{2}$, i.e. $\tan 22\frac{1}{2}° = \sqrt{2}-1$

n The other root of the quadratic equation is $-1-\sqrt{2}$; can you suggest an angle the tangent of which is $-1-\sqrt{2}$? Can you adjust the working above to find the exact value of $\tan 67\frac{1}{2}°$?

Exercise 32

(1) Solve, for $0° < \theta < 360°$, the equation $\cot^2\theta + \csc^2\theta + 3\csc\theta = 19$.

(2) Given that α is the acute angle such that $\sin\alpha = \frac{8}{17}$, find the exact value of

   (i) $\sin 2\alpha$ **(ii)** $\cos(\alpha + 60°)$

(3) Express $\sqrt{5}\cos\theta - \sqrt{2}\sin\theta$ in the form $R\cos(\theta + \alpha)$, where $R > 0$ and $0° < \alpha < 90°$. Hence solve, for $-180° < \theta < 180°$, the equation $\sqrt{5}\cos\theta - \sqrt{2}\sin\theta = 2$.

(4) **(i)** Use an identity for $\cos 2\theta$ to show that $\cos 15° = \frac{1}{2}\sqrt{2+\sqrt{3}}$.

   (ii) Expand $\cos(45° - 30°)$ to find the exact value of $\cos 15°$ in a different form.

(5) Prove that $\dfrac{\cos 4\theta \sin 2\theta}{\sin\theta} \equiv 16\cos^5\theta - 16\cos^3\theta + 2\cos\theta$.

(6) Given that $\tan\phi = \frac{1}{2}$, find the exact value of $\tan 4\phi$.

Numerical solution of equations

Number of real roots

Consider the equation $x^3 + 4x - 25 = 0$. There is no easy direct way to solve it algebraically, but there are various numerical methods that can be used. First, it would be helpful to know how many real roots the equation has; sketch-graphs can be used to determine this.

The equation can be rewritten as $x^3 = 25 - 4x$. Sketching the graphs of $y = x^3$ and $y = 25 - 4x$ on the same set of axes gives:

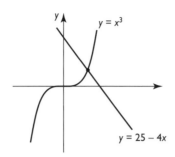

The two graphs meet only once, i.e. there is only one value of x for which x^3 and $25 - 4x$ have the same value. The equation therefore has one real root.

Worked example

Find the number of real roots of the equation $3e^{2x} + x^2 - 10 = 0$.

Solution

n First we must decide how to rearrange the equation so that we get expressions on the left- and right-hand sides whose graphs we can easily sketch.

Rearranging gives $3e^{2x} = 10 - x^2$

Sketching graphs of $y = 3e^{2x}$ and $y = 10 - x^2$:

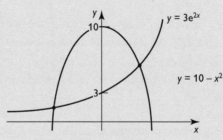

Clearly, these graphs meet twice and so the equation has two real roots.

n In drawing the sketches, we must pay attention to the relative positions of the two graphs. Here, one curve crosses the y-axis at $(0, 10)$ and the other at $(0, 3)$.

Exercise 33

(1) Use sketch-graphs to determine the number of real roots of each of the following equations.

  (i) $x^4 + 5x - 11 = 0$

  (ii) $3\ln x + x = 8$

  (iii) $10e^{-2x} = (x + 3)(x - 4)^2$

Location of roots

We return to the equation $x^3 + 4x - 25 = 0$ studied earlier. Before attempting to find the root, it is advisable to know its approximate location. Define $f(x) = x^3 + 4x - 25$; this means that we are looking for the value of x for which $f(x) = 0$. Substituting various values of x gives $f(0) = -25$, $f(1) = -20$, $f(2) = -9$ and $f(3) = 14$.

As the value of x goes from 2 to 3, the value of $f(x)$ changes from negative to positive. So, between $x = 2$ and $x = 3$, the corresponding graph must cross the x-axis:

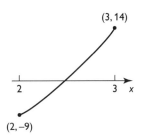

Hence the root is between 2 and 3.

We could carry out further calculations, for example: $f(2.2) = -5.552$, $f(2.4) = -1.576$ and $f(2.6) = 2.976$.

Now the sign change indicates that the root is between 2.4 and 2.6.

Worked example
Find the pair of consecutive integers between which the positive root of the equation $3e^x + x^2 = 75$ lies.

Solution
⚠ The first step is to rearrange the equation to the form $... = 0$, because only with the equation in this form is it valid to search for a sign change.

Define $f(x) = 3e^x + x^2 - 75$ and substitute integers until a sign change is noted:
$f(0) = -72$, $f(1) = -65.8$, $f(2) = -48.8$, $f(3) = -5.7$, $f(4) = 104.8$

The sign change in $f(x)$ from -5.7 to 104.8 indicates that the positive root lies between $x = 3$ and $x = 4$

⚠ There is an assumption, in recognising the significance of the sign change, that the curve is *continuous* between the two x values.

Exercise 34
(1) Find by calculation the pair of consecutive integers between which the root of the equation $x^5 + 8x - 40 = 0$ lies.

(2) Show by calculation that the root of the equation $8\ln(x + 1) = (3 + x)(5 - x)$ lies between 3 and 3.5.

Finding roots

One possible rearrangement of the equation $x^3 + 4x - 25 = 0$ is $x = \sqrt[3]{25 - 4x}$. Consider the associated iterative formula $x_{n+1} = \sqrt[3]{25 - 4x_n}$. We established earlier that the root of $x^3 + 4x - 25 = 0$ is between 2.4 and 2.6. Taking $x_1 = 2.5$ and iteratively applying the formula $x_{n+1} = \sqrt[3]{25 - 4x_n}$ leads to the following sequence:

$x_2 = 2.466212$, $x_3 = 2.473597$, $x_4 = 2.471987$, $x_5 = 2.472338$, $x_6 = 2.472261$, $x_7 = 2.472278$, $x_8 = 2.472274,...$

The terms of the sequence are converging and, correct to 4 decimal places, the root is 2.4723.

Not all rearrangements will give rise to an iterative formula that generates terms which converge to the root. For example, check what happens if you use $x_1 = 2.5$ and $x_{n+1} = \frac{1}{4}(25 - x_n^3)$ instead. In an examination question, it will always be made clear what iterative formula is to be used.

Worked example

(i) State the equation which the iterative formula $x_{n+1} = \frac{1}{2}\ln\frac{20 - 7x_n}{3}$ with a suitable starting value is designed to solve.

(ii) Use $x_1 = 1$ to find the root of the equation correct to 3 decimal places.

Solution

n If the sequence converges, then eventually the value of X which is put in (X_n) will be the same as the value obtained (X_{n+1}); so the equation involved is found by replacing both X_{n+1} and X_n by X.

(i) The equation is $x = \frac{1}{2}\ln\frac{20 - 7x}{3}$

which can be rearranged to $2x = \ln\frac{20 - 7x}{3}$, then $e^{2x} = \frac{20 - 7x}{3}$ and hence $3e^{2x} + 7x - 20 = 0$

(ii) Applying the iterative formula:

$1 \to 0.73317 \to 0.80029 \to 0.78424 \to 0.78812 \to 0.78719 \to 0.78741$

So the root is 0.787 correct to 3 decimal places.

n If your calculator has an ANS button, it is easy to produce the sequence of values. First enter the starting value 1. The formula $\frac{1}{2}\ln((20 - 7\text{ANS}) \div 3)$ with the ENTER or EXECUTE button pressed repeatedly will then generate successive terms in the sequence. You must continue until you are certain that the output values have settled down to the requested number of significant figures or decimal places. In answering an examination question, you should provide details of all the terms produced.

Exercise 35

(1) (i) Use a diagram to show that the equation $x^7 = x^2 + 9$ has exactly one real root.

   (ii) Show by calculation that the root lies between 1 and 2.

   (iii) Use an iteration process based on the equation $x = \sqrt[7]{x^2 + 9}$, with a suitable starting value, to find the root correct to 4 decimal places.

(2) It is given that the equation $(x - 2)^3 = 6 - 2e^{-3x}$ has two real roots α and β, where $\alpha < 0$ and $\beta > 0$.

   (i) Confirm by means of appropriate graphs that the equation has exactly two real roots.

   (ii) Find by calculation the pair of consecutive integers between which α lies and the pair of consecutive integers between which β lies.

   (iii) The equation $(x - 2)^3 = 6 - 2e^{-3x}$ can be rewritten as either $x = -\frac{1}{3}\ln[3 - \frac{1}{2}(x - 2)^3]$ or as $x = 2 + \sqrt[3]{6 - 2e^{-3x}}$. Use an appropriate iterative formula to find the values of α and β correct to 4 decimal places.

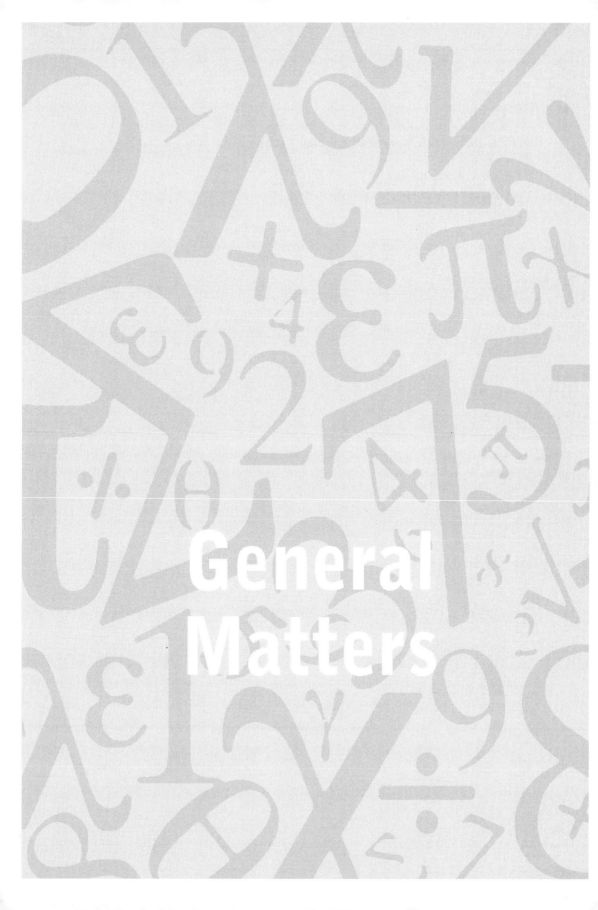

General
Matters

This section looks at three important general aspects of preparing to take the
A2 Module 4723: Core Mathematics 3 (C3) examination:

- calculators
- the formulae booklet
- the relevance of previous mathematical knowledge

general matters

Calculators

In the C3 examination, you are allowed to use a scientific calculator, a graphical calculator, or both; however, calculators that possess computer algebra functions are not permitted. You are expected to be able to use your calculator accurately and efficiently, so you need to be completely familiar with what your calculator is capable of doing. Can it generate the terms of a sequence defined by an iterative formula? Can it provide exact answers?

A significant topic in C3 is trigonometry, and it is important to be alert to whether the calculator should operate in DEGREE mode or in RADIAN mode.

Although it is not essential to have a graphical calculator, careful and informed use of one should help to enhance your understanding of mathematics. When using such a calculator to draw a graph, extra care needs to be taken in setting the scales on the axes. Consider the graph of $y = e^x - 1$. A sketch needs to show key features such as the fact that the curve passes through the origin, that the line $y = -1$ is an asymptote and that the curve has an increasing, positive gradient:

But, with inappropriately set scales, it is all too easy for a calculator to display a graph like the one below which, in an examination question, would certainly not earn all the credit.

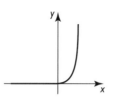

Exercise 36

(1) A sequence is defined by $u_1 = 3$, $u_{n+1} = \sqrt{2u_n + 5}$. Find the values of u_2, u_3, u_4, u_5 and u_6, giving each correct to 4 significant figures.

(2) Solve, for $0 < x < 2\pi$, the equation $4\operatorname{cosec} x = 15$.

(3) Solve, for $0 < \theta < 360°$, the equation $4\cot\theta = 15$.

(4) Sketch the curve with equation $y = 4\ln(3x - 1)$.

(5) Find the exact value of $\displaystyle\int_2^5 \frac{8}{2x - 1}\,dx$.

Formulae booklet

The *List of Formulae* booklet (MF1) is an important resource in the C3 examination. It is essential that you become very familiar with what is contained in this booklet and how you can make use of it during the examination. The same booklet is provided in all of the OCR A-level mathematics examinations, so it includes formulae and statistical tables for all the different AS and A2 units. It is a good idea to practise using the booklet while preparing for the examination, so that you learn where to find quickly any formula that you might need. You can obtain a copy of the booklet from the **www.ocr.org.uk** website.

The booklet contains many formulae that you will have met while studying the modules C1, C2 and C3. The formulae that are of most relevance to C3 are:

- $e^{x \ln a} = a^x$
- $\sin(A \pm B) \equiv \sin A \cos B \pm \cos A \sin B$
- $\cos(A \pm B) \equiv \cos A \cos B \mp \sin A \sin B$
- $\tan(A \pm B) \equiv \dfrac{\tan A \pm \tan B}{1 \mp \tan A \tan B} \qquad (A \pm B \neq (k + \frac{1}{2})\pi)$
- if $y = \dfrac{f(x)}{g(x)}$, then $\dfrac{dy}{dx} = \dfrac{f'(x)g(x) - f(x)g'(x)}{\{g(x)\}^2}$
- Simpson's rule:

$$\int_a^b y \, dx \approx \tfrac{1}{3} h\{(y_0 + y_n) + 4(y_1 + y_3 + \dots + y_{n-1}) + 2(y_2 + y_4 + \dots + y_{n-2})\},$$

where $h = \dfrac{b-a}{n}$ and n is even

Not all the formulae and results that you might need for the C3 examination are listed in the booklet. Therefore, you must know these other formulae and results by heart; the following exercise invites you to check whether you can remember them.

Exercise 37

(1) Complete the identity $\sin 2A \equiv \dots$

(2) Given $y = \ln(ax + b)$, find $\dfrac{dy}{dx}$.

(3) Write down the condition for the quadratic equation $ax^2 + bx + c = 0$ to have no real roots.

(4) State the exact value of $\operatorname{cosec} 60°$.

(5) Write down an identity linking $\tan^2 \theta$ and $\sec^2 \theta$.

(6) Given $y = e^{kx}$, find $\dfrac{dy}{dx}$.

(7) Write down the formula for the volume of the solid produced when a curve is rotated completely about the x-axis.

(8) Express $2 \ln a + \ln b - \ln c$ as a single logarithm.

(9) Complete the identity $\tan 2A \equiv \dots$

(10) Write down the product rule for differentiation.

(11) State the discriminant of $ax^2 + bx + c$.

(12) State the exact value of $\cot\frac{5}{6}\pi$.

(13) Write down an identity linking $\csc^2\theta$ and $\cot^2\theta$.

(14) Write down the formula for the volume of the solid produced when a curve is rotated completely about the y-axis.

(15) Find $\int e^{kx}dx$.

(16) Express $\cos 2\theta$ in terms of $\cos\theta$.

(17) State the exact value of $\sec\frac{5}{4}\pi$.

(18) Find $\displaystyle\int\frac{1}{ax+b}dx$.

(19) Given $Q = 2\log_{10}(t + 1)$, express t in terms of Q.

(20) Sketch the graph of $y = 4^{x+2}$.

Synoptic assessment

In mathematics, the very nature of the subject means that the requirement for synoptic assessment is met automatically. In answering C3 questions, you will often be using results and techniques that you studied earlier in C1 or C2. The following exercise consists of C3 questions that depend in part on ideas from preceding modules.

Exercise 38

(1) Using the substitution $u = e^{\frac{1}{2}x}$ or otherwise, solve the equation $2e^x + 5\,e^{\frac{1}{2}x} = 12$.

(2) Find the equation of the normal to the curve $y = \dfrac{9}{2x + 5}$ at the point $(-1, 3)$.

(3) By completing the square, determine the range of the function f defined by $f(x) = 2x^2 - 6x - 1$ for all real values of x.

(4) Show that $\displaystyle\int_0^{18}\frac{1}{\sqrt{4x + 3}}dx = 2\sqrt{3}$.

(5) Solve the inequality $|5x - 1| < 2|x|$.

(6) A curve has $\dfrac{dy}{dx} = 80(3 - 2x)^7$ and passes through the point $(1, 15)$. Find the equation of the curve.

(7) The convergent geometric series $1 + 2e^{-x} + 4e^{-2x} + \ldots$ has sum to infinity equal to $\frac{5}{4}$. Find the value of x.

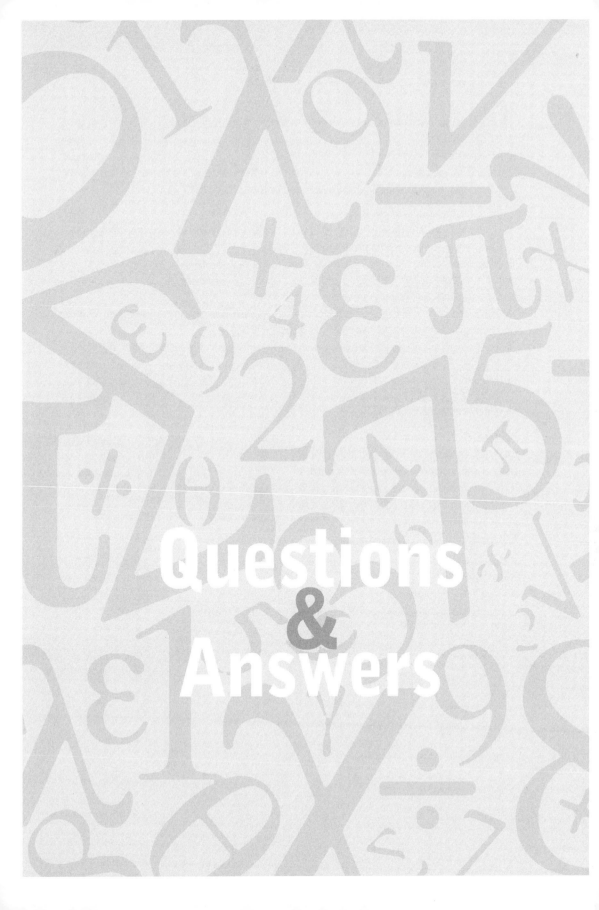

Questions
&
Answers

This section consists of three practice C3 examination papers, each designed to be of the same style and standard as an actual C3 examination paper. Each paper is worth a total of 72 marks, and the time allowed is 1 hour 30 minutes. The best approach is to set aside this time and work steadily through a paper. Practising in this way should help significantly when you come to take the examination.

Remember the following points:

- The early questions in the paper should be straightforward, whereas later questions may contain some challenging aspects.
- The best strategy is to work through the questions in order.
- Although these are practice papers, answer them with care — you do not want to develop bad habits.
- Have the *List of Formulae* booklet to hand as you attempt these papers; it will be an important aid as you answer the questions.
- In the examination, you must present your solutions clearly and fully, otherwise you risk losing marks. Do the same with these practice papers.
- Make sure that your calculator is in good working order. Also make sure that you have it set to the correct mode; in this module, you will sometimes need it in DEGREE mode and sometimes in RADIAN mode.

For each of the first two practice papers, there is a set of hints and suggestions, question by question. Consult this only if you are really stuck; it is much more beneficial if you can sort out problems without looking at the hints. There will not be any hints available to you in the examination, and dealing with problems is an important examination technique that needs practice.

Examiner's comments

Solutions to the questions are provided. These are followed by examiner's comments indicated by the icon 𝑒, which point out various issues such as common errors, tricky points, alternative methods etc. Once you have completed a paper, check your solutions and read carefully through the comments.

Specimen Paper 1

Question 1

Find

(i) $\int (5x + 7)^4 dx$.   (2 marks)

(ii) $\int (5x + 7)^{-1} dx$.   (2 marks)

Question 2

Find the equation of the tangent to the curve $y = (x^3 + 9)^{\frac{1}{2}}$ at the point $(3, 6)$, giving your answer in the form $ax + by + c = 0$ where a, b and c are integers.   (5 marks)

Question 3

Given that the value of x is such that $|x - 5| = |x + 9|$, find the value of $|3x - 7|$.   (5 marks)

Question 4

The functions f and g are defined for all real values of x by $f(x) = 2x^2 + 3$ and $g(x) = 1 - 4x$.

(i) State the range of f.   (1 mark)

(ii) Evaluate $fg(-2)$.   (2 marks)

(iii) Find an expression for $g^{-1}(x)$.   (3 marks)

Question 5

(a) Given that $3 \cos 2\alpha = 13 \cos \alpha + 2$, find the exact value of $\cos \alpha$.   (4 marks)

(b) Given that $\tan 2\beta = 4 \cot \beta$, find the possible exact values of $\tan \beta$.   (4 marks)

Question 6

Find the exact gradient of each of the following curves at the point for which $x = 2$:

(i) $y = 4 \ln(2x + 1)$ (3 marks)

(ii) $y = \dfrac{5x + 3}{4x - 1}$ (4 marks)

Question 7

(i) By drawing sketches of $y = e^{2x}$ and $y = 150 - 6x^2$ on the same diagram, show that the equation $e^{2x} = 150 - 6x^2$ has exactly two real roots.   (3 marks)

(ii) Use your diagram to show that one root is close to -5.   (2 marks)

(iii) Show by calculation that the second root is between 2 and 2.5.   (3 marks)

(iv) Use an iteration process, based on the equation $x = \frac{1}{2} \ln(150 - 6x^2)$ with a suitable starting value, to find the second root correct to 3 decimal places.   (4 marks)

Question 8

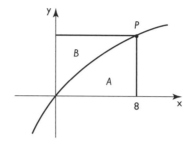

The diagram shows the curve with equation $y = 2\ln(x + 1)$. The point P on the curve has x-coordinate 8. The region A is bounded by the curve and the lines $x = 8$ and $y = 0$. The region B is bounded by the curve, the line $x = 0$ and the line through P parallel to the x-axis.

(i) The curve $y = \ln x$ can be transformed to the curve $y = 2\ln(x + 1)$ by a pair of transformations. Give details of these transformations. (3 marks)

(ii) Use Simpson's rule with four strips to find an approximation to the area of A. (4 marks)

(iii) Show that the area of region B is given by $\int_0^h (e^{\frac{1}{2}y} - 1)dy$, where the exact value of h is to be stated. Hence find the exact area of region B. (5 marks)

Question 9

(i) Express $\sqrt{3}\sin\theta - \sqrt{2}\cos\theta$ in the form $R\sin(\theta - \alpha)$, where $R > 0$ and $0 < \alpha < \frac{1}{2}\pi$. (3 marks)

(ii) Hence

(a) solve, for $0 \le \theta \le 2\pi$, the equation $\sqrt{3}\sin\theta - \sqrt{2}\cos\theta = 1$ (3 marks)

(b) find the greatest possible value, as θ varies, of $\dfrac{5}{\sqrt{3}\sin\theta - \sqrt{2}\cos\theta + \sqrt{20}}$ (3 marks)

(c) find the greatest possible value, as θ varies, of $\left(\sqrt{3}\sin\frac{1}{2}\theta - \sqrt{2}\cos\frac{1}{2}\theta + 1\right)^2$, and determine a value of θ for which that greatest value occurs (4 marks)

Hints and suggestions

Question 1

Part (i) is an easy start to the paper; note, however, that $\frac{1}{5}(5x + 7)^5 + c$ is wrong. Part (ii) involves a natural logarithm, of course.

Question 2

Differentiation is needed and this must involve the chain rule. Make sure you obtain a numerical value for the gradient. A little manipulation is needed at the end to put the equation into the required form.

Question 3

This is really a two-stage question. First, solve the equation $|x - 5| = |x + 9|$. With such a straightforward equation, it is easy (and certainly worth the effort) to check that your answer (there is only one answer) is indeed correct. The second stage of the solution is to substitute the value you obtained into $|3x - 7|$; here you will need to remember what the modulus sign means.

Question 4

Part (i) says 'State...', so expect to be able to write the answer down immediately. Recall that the range is the set of resulting y-values. To find fg(–2) in part (ii), remember it is the function on the right, namely g, that is applied first. In part (iii), start by writing down $y = 1 - 4x$; then rearrange this to the form $x = ...$

Question 5

This question is assessing, in part, whether you are able to recall the double-angle formulae. If you cannot do so, you can always refer to the *List of Formulae* and, for example, produce the identity for $\cos 2\alpha$ by treating it as $\cos(\alpha + \alpha)$. Part (a) leads to a quadratic equation in $\cos \alpha$ but, as the question indicates, only one of the two roots of the quadratic is a valid answer. In part (b), you will need the identity for $\tan 2\beta$ as well as the definition of $\cot \beta$. Do not make the mistake of thinking that $4 \cot \beta = \dfrac{1}{4 \tan \beta}$.

Question 6

This is a routine test of differentiation techniques. Note that in part (i), the derivative is not $\dfrac{4}{2x+1}$. Use brackets appropriately in part (ii) when you apply the quotient rule — it is easy to make sign errors here.

Question 7

Both sketch-graphs should be familiar; be sure to show the key features of each graph. A pair of good sketches should show one intersection very close to the x-axis, a fact which leads to the result in part (ii). Make your explanation as clear as you can. For part (iii), before searching for a change of sign you must rearrange the equation given in part (i). Show all the details in part (iv). It is the final answer that is required to be correct to 3 decimal places, so you should show at least 4 decimal places in the sequence of values obtained from the iterative formula.

Question 8

One transformation in part (i) is a translation and the other is a stretch; provide the details of each precisely. In part (iii), remember that the area of a region bounded between a curve and the y-axis is given by $\int_c^d x\,dy$, and x must be expressed in terms of y before any integration is attempted.

Question 9

Note that the angles here are in radians. There are two values to be found in part (ii)(a). The key to the final two parts is the realisation that the value of $\sin(\theta - \alpha)$ must lie between -1 and $+1$ (inclusive).

Solutions

Question 1

(i) $\int (5x+7)^4\,dx = \frac{1}{25}(5x+7)^5 + c$

(ii) $\int (5x+7)^{-1}\,dx = \frac{1}{5}\ln|5x+7| + c$

 This is an easy start to the paper for most candidates, although many omit the arbitrary constant of integration. In part (ii), not all candidates recognise this as an integral that involves a logarithm, and solutions including $(5x + 7)^0$ sometimes occur.

Question 2

$y = (x^3+9)^{\frac{1}{2}} \Rightarrow \dfrac{dy}{dx} = \frac{1}{2}(x^3+9)^{-\frac{1}{2}}\cdot 3x^2 = \frac{3}{2}x^2(x^3+9)^{-\frac{1}{2}} = \dfrac{3x^2}{2\sqrt{x^3+9}}$

When $x = 3$, $\dfrac{dy}{dx} = \dfrac{3\times 9}{2\sqrt{36}} = \dfrac{27}{12} = \dfrac{9}{4}$

Equation of tangent is $y - 6 = \frac{9}{4}\,(x - 3)$

Multiplying by 4 gives $4y - 24 = 9x - 27$

Rearranging to requested form, the equation is $9x - 4y - 3 = 0$

 The chain rule is needed here but not all candidates realise this. Some candidates do not appreciate that a numerical value for the gradient is required and propose an incorrect equation such as $y - 6 = \frac{3}{2}x^2(x^3 + 9)^{-\frac{1}{2}}(x - 3)$ for the tangent. Many candidates lose the final mark for not presenting the answer in the requested form.

Question 3

$|x - 5| = |x + 9| \Rightarrow x - 5 = \pm(x + 9)$, leading to $x - 5 = x + 9$ or $x - 5 = -x - 9$

$x - 5 = x + 9$ has no solution, while $x - 5 = -x - 9$ gives $2x = -4$ and hence $x = -2$

Substituting $x = -2$ into $|3x - 7|$ gives $|3 \times (-2) - 7| = |-13| = 13$

 Most candidates solve the modulus equation correctly, but there is more uncertainty when it comes to finding the value of $|3x - 7|$. In particular, some candidates are unsure of what $|-13|$ means, and answers such as 169 are common.

Question 4

(i) Range of f is f(x) ≥ 3

(ii) fg(−2) = f(9) = 165

(iii) Writing $y = 1 - 4x$ and rearranging, we get $4x = 1 - y$ and hence $x = \frac{1}{4}(1 - y)$

Therefore $g^{-1}(x) = \frac{1}{4}(1 - x)$

✎ Not all candidates understand what is meant by range, so many are unable to write down the correct answer to part (i). More assured work is evident in parts (ii) and (iii), and the vast majority of candidates answer these well.

Question 5

(a) $3\cos 2\alpha = 13\cos\alpha + 2$

$\Rightarrow 3(2\cos^2\alpha - 1) = 13\cos\alpha + 2$

$\Rightarrow 6\cos^2\alpha - 13\cos\alpha - 5 = 0$

Factorising gives $(3\cos\alpha + 1)(2\cos\alpha - 5) = 0$

so $\cos\alpha = -\frac{1}{3}$ (since $\cos\alpha = \frac{5}{2}$, from the second bracket, is impossible)

(b) $\tan 2\beta = 4\cot\beta$

$\Rightarrow \dfrac{2\tan\beta}{1 - \tan^2\beta} = \dfrac{4}{\tan\beta}$

$\Rightarrow 2\tan^2\beta = 4(1 - \tan^2\beta)$

i.e. $6\tan^2\beta = 4$, $\tan^2\beta = \frac{2}{3}$

so $\tan\beta = \pm\sqrt{\frac{2}{3}} = \pm\frac{1}{3}\sqrt{6}$

✎ The double-angle identities do not appear in the *List of Formulae* and many candidates are unsure of them. Some candidates do not discard the value $\frac{5}{2}$ in part (a), whereas a common error in part (b) is a failure to realise that $\tan^2\beta = \frac{2}{3}$ leads to a negative as well as a positive value. Two completely correct solutions indicate work of at least grade B standard.

Question 6

(i) $y = 4\ln(2x + 1) \Rightarrow \dfrac{dy}{dx} = \dfrac{8}{2x+1}$

When $x = 2$, $\dfrac{dy}{dx} = \dfrac{8}{5}$

(ii) $y = \dfrac{5x+3}{4x-1} \Rightarrow \dfrac{dy}{dx} = \dfrac{5(4x-1) - 4(5x+3)}{(4x-1)^2} = -\dfrac{17}{(4x-1)^2}$

When $x = 2$, $\dfrac{dy}{dx} = -\dfrac{17}{49}$

✎ These requests usually present no difficulties. The errors that occur most often are giving the derivative as $\dfrac{4}{2x+1}$ in part (i) and making sign mistakes when applying the quotient rule in part (ii).

Question 7

(i)

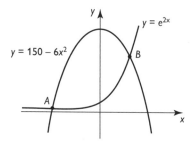

The curves meet at two points; the x-values associated with the intersection points A and B are the two roots.

(ii) e^{2x} is small for negative x-values, so the root corresponding to point A in the diagram should be close to the negative solution of $150 - 6x^2 = 0$

Solving $150 - 6x^2 = 0$ gives $6x^2 = 150$, $x^2 = 25$ and hence $x = \pm 5$

Thus the root associated with point A is close to -5

(iii) Putting $f(x) = e^{2x} + 6x^2 - 150$, the roots correspond to solutions of $f(x) = 0$

Evaluating, $f(2) = e^4 + 24 - 150 = -71.4$ and $f(2.5) = e^5 + 37.5 - 150 = 35.9$

Since $f(2) < 0$ but $f(2.5) > 0$, a root lies between 2 and 2.5

(iv) Using the iterative formula $x_{n+1} = \frac{1}{2}\ln(150 - 6x_n^2)$ with starting value 2.25:

$2.25 \rightarrow 2.39218 \rightarrow 2.37535 \rightarrow 2.37743 \rightarrow 2.37717$

The second root is 2.377 (correct to 3 decimal places)

Plotted graphs are not expected in part (i); indeed, with y-intercepts of $(0, 150)$ and $(0, 1)$ for the two curves, it is difficult to draw them to scale and show their essential features at the same time. With part (ii) depending crucially on the nature of the sketches for negative x, it is important to have a diagram that shows both curves clearly near $x = -5$. Part (iii) causes difficulties and many candidates do not understand that an equation must be in the $f(x) = 0$ form to justify searching for a sign change. Other approaches are possible for part (iii), but these require careful explanations and formulating such explanations is challenging for many candidates. Part (iv) is generally answered well, although not all candidates provide sufficient evidence to support their concluding value.

Question 8

(i) A stretch by scale factor 2 in the y-direction and translation by 1 unit in the negative x-direction (in either order)

(ii) Area of $A \approx \frac{1}{3} \times 2(2\ln 1 + 4 \times 2\ln 3 + 2 \times 2\ln 5 + 4 \times 2\ln 7 + 2\ln 9) = 23.5$ (to 3 significant figures)

(iii) Rearranging $y = 2\ln(x + 1)$ to make x the subject gives $\frac{1}{2}y = \ln(x + 1)$, $e^{\frac{1}{2}y} = x + 1$,

and hence $x = e^{\frac{1}{2}y} - 1$

$$\text{Area of } B = \int_0^{2\ln 9} x\,dy = \int_0^{2\ln 9} (e^{\frac{1}{2}y} - 1)\,dy = \left[2e^{\frac{1}{2}y} - y\right]_0^{2\ln 9}$$

$$= 2e^{\ln 9} - 2\ln 9 - (2 - 0) = 18 - 2\ln 9 - 2 = 16 - 2\ln 9$$

✎ To earn all 3 marks in part (i), the details of both transformations must be precise and use appropriate terminology; full credit will not be given if words such as 'move' or 'shift' are used. Part (ii) is answered well, although some candidates are uncertain about what coefficients to use with the y-values, with some mistakenly adopting a 1, 2, 4, 2, 1 pattern. The fact that the area between a curve and the y-axis is given by $\int_c^d x\,dy$ is not widely known, so candidates often have difficulty establishing the result in part (iii). Candidates scoring well on this question are exhibiting work of grade A standard.

Question 9

(i) $\sqrt{3}\sin - \sqrt{2}\cos\theta = R\sin(\theta - \alpha) \equiv R\sin\theta\cos\alpha - R\cos\theta\sin\alpha$

So we need $R\cos\alpha = \sqrt{3}$ and $R\sin\alpha = \sqrt{2}$

Squaring and adding, $R^2 = 3 + 2$, so $R = \sqrt{5}$

Dividing, $\tan\alpha = \dfrac{\sqrt{2}}{\sqrt{3}}$, giving $\alpha = 0.68472$ (radians)

Hence $\sqrt{3}\sin\theta - \sqrt{2}\cos\theta \equiv \sqrt{5}\sin(\theta - 0.68472)$

(ii) (a) $\sqrt{3}\sin\theta - \sqrt{2}\cos\theta = 1$ is equivalent to $\sqrt{5}\sin(\theta - 0.68472) = 1$

i.e. $\sin(\theta - 0.68472) = \dfrac{1}{\sqrt{5}}$

This yields $\theta - 0.68472 = 0.46365$ or $\pi - 0.46365$

giving $\theta = 1.15, 3.36$

(ii) (b) $\dfrac{5}{\sqrt{3}\sin\theta - \sqrt{2}\cos\theta + \sqrt{20}} = \dfrac{5}{\sqrt{5}\sin(\theta - 0.68472) + \sqrt{20}}$

Value is greatest when the denominator is as small as possible.

The least value of $\sin(\theta - 0.68472)$ is -1, so the greatest value of the given

expression is $\dfrac{5}{\sqrt{5}\times(-1)+\sqrt{20}} = \dfrac{5}{-\sqrt{5}+2\sqrt{5}} = \dfrac{5}{\sqrt{5}} = \sqrt{5}$

(ii) (c) $\left(\sqrt{3}\sin\frac{1}{2}\theta - \sqrt{2}\cos\frac{1}{2}\theta + 1\right)^2 = \left[\sqrt{5}\sin(\frac{1}{2}\theta - 0.68472) + 1\right]^2$

The greatest possible value of $\sin(\frac{1}{2}\theta - 0.68472)$ is 1

hence the greatest value of $\left[\sqrt{5}\sin(\frac{1}{2}\theta - 0.68472) + 1\right]^2$ is

$\left[\sqrt{5} + 1\right]^2 = 5 + 2\sqrt{5} + 1 = 6 + 2\sqrt{5}$

This greatest value occurs when $\sin(\frac{1}{2}\theta - 0.68472) = 1$,

i.e. when $\frac{1}{2}\theta - 0.68472 = \frac{1}{2}\pi$, or $\theta = 4.51$

This proves to be quite a challenging question. Even the first two parts present problems because radians are involved — many candidates are not comfortable in dealing with radians and are unsure, in part (i), whether to give α as 0.6847 or as 0.6847π. Finding the second value of θ in part (ii)(a) is a problem for many who do not appreciate that the correct procedure is to first determine a second value for $\theta - 0.6847$. The key to the final two parts is the realisation that sine takes values between -1 and $+1$. Candidates applying this idea successfully in (ii)(b) and (ii)(c) are showing pleasing mathematical insight, which is certainly indicative of grade A standard.

Specimen Paper 2

Question 1

Find the coordinates of the point on the curve $y = (3x - 2)^4$ at which the gradient is 96. (4 marks)

Question 2

The sequence $x_1, x_2, x_3,...$ is defined by $x_1 = 1, x_{n+1} = (2 + e^{-2x_n})^{\frac{1}{3}}$ and converges to the value α.

(i) Find the value of α correct to 3 decimal places. (4 marks)

(ii) Find an equation of which α is a root. (2 marks)

Question 3

Solve the inequality $|3x - 5| \leq |2x + 1|$. (6 marks)

Question 4

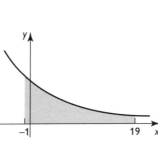

The diagram shows the curve $y = \dfrac{1}{\sqrt{2x + 7}}$. The shaded region is bounded by the curve and the lines $x = -1, x = 19$ and $y = 0$. The shaded region is rotated completely about the x-axis. Find the exact volume of the solid produced, giving your answer in simplified form. (6 marks)

Question 5

It is given that θ is the acute angle such that $\cos\theta = \frac{1}{3}$.

(i) Find the exact value of $\sin 2\theta$. (4 marks)

(ii) Find the value of $\tan(\theta - 45°)$, giving your answer in the form $a + b\sqrt{2}$. (4 marks)

Question 6

Gas is being pumped into a spherical balloon at a constant rate of 0.65 cubic metres per minute. Find the rate at which the radius is increasing at the instant when the radius of the balloon is 2 metres. (5 marks)

Question 7

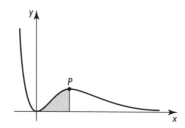

The diagram shows the curve with equation $y = x^2 e^{-3x}$.

(i) Use differentiation to show that the curve has a stationary point at the origin and find the exact coordinates of P, the other stationary point. (6 marks)

(ii) The region shaded in the diagram is bounded by the curve, the line $y = 0$ and the line through P parallel to the y-axis. Use Simpson's rule with two strips to find an approximation to the area of the shaded region. (3 marks)

Question 8

(i) Prove that $\tan^2 \theta(\sec^2 \theta + \cosec^2 \theta) \equiv \sec^4 \theta$. (3 marks)

(ii) Hence

(a) solve, for $0° \le \alpha \le 360°$, the equation $5 \tan^2 \alpha(\sec^2 \alpha + \cosec^2 \alpha) = 11$; (3 marks)

(b) state the values of the constant k for which there are no values of β that satisfy the equation $\tan^2 \beta(\sec^2 \beta + \cosec^2 \beta) = k$. (3 marks)

Question 9

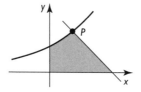

The diagram shows the curve with equation $y = 2e^{\frac{1}{4}x}$. The point P on the curve has x-coordinate ln 16. The shaded region is bounded by the curve, the axes and the normal to the curve at P. Show that the area of the shaded region is 16 square units. (9 marks)

Question 10

The function f is defined for all real values of x by $f(x) = x^2 - 6x + c$.

(i) Given that the range of f is $f(x) \ge 4$, find the value of c. (3 marks)

(ii) Given instead that $f(x) = |f(x)|$ for all real values of x, find the possible values of c. (3 marks)

(iii) Show that there are no values of c for which $ff(4) = 0$. (4 marks)

Hints and suggestions

Question 1

The derivative needs to be equated to 96 and the resulting equation solved. The question does ask for the 'coordinates', so you have to find the y-value of the point too.

Question 2

Be sure to show the terms of the sequence. With these, you should also write down more than 3 decimal places; but your final value of α should show precisely 3 decimal places. The equation in part (ii) is obtained by replacing both x_{n+1} and x_n by x.

Question 3

This is a routine request which ought to be familiar. If you take the approach of squaring both sides, remember that care is needed when solving the resulting quadratic inequality — draw a sketch to establish the required set of values.

Question 4

The formula for a volume of revolution is not given in the *List of Formulae*, so you must memorise it. If you are not sure of the formula, look it up now and commit it to memory. The integration here will involve a natural logarithm.

Question 5

The first step is to find the exact value of $\sin\theta$, using either the identity $\sin^2\theta + \cos^2\theta \equiv 1$ or an appropriate right-angled triangle. The identity for $\sin 2\theta$ is not in the *List of Formulae* and is therefore something you should remember. Part (ii) requires the identity for $\tan(A - B)$; this is included in the *List of Formulae*.

Question 6

This is a question about connected rates of change. You are expected to know that the formula for the volume of a sphere is $V = \frac{4}{3}\pi r^3$. Write down accurately what the question tells you. One such statement is $\dfrac{dV}{dt} = 0.65$, and it is the value of $\dfrac{dr}{dt}$ that you are asked to find.

Question 7

The differentiation in part (i) requires the product rule. Equating the derivative to zero to find the x-coordinates of the stationary points requires careful work — both to confirm that there is indeed a stationary point at the origin and to find the x-coordinate of P. In part (ii), use of Simpson's rule with just two strips means that the formula you need is $\frac{1}{3}h(y_0 + 4y_1 + y_2)$.

Question 8

It may not be immediately clear how to proceed in proving the identity in (i), but a start has to be made and the most promising approach might be to express the left-hand side in terms of $\sin\theta$ and $\cos\theta$, and then expect that appropriate simplification will lead to $\sec^4\theta$. Note that the mark allocation is 3, which is an indication that the proof does not need to be long. (An alternative neat proof involves expressing the left-hand side in terms of $\tan\theta$.) The requests in part (ii) start with the word 'Hence', so you can expect to rely on the identity from part (i). For (ii)(b), you need to consider the values that $\sec\theta$, and hence $\sec^4\theta$, can take.

Question 9

This question is not structured into parts, so you must plan the strategy needed to construct the solution. Also note that the answer is given; your solution will therefore need to be suitably detailed to convince the examiner that you really understand what you are doing.

Question 10

For part (i), it will help if the expression for $f(x)$ is written in completed square form. There is only a single correct value for c, not a range of values. For part (ii), recall how the graph of $y = |f(x)|$ is drawn; what is the implication if the graph of $y = f(x)$ is the same as the graph of $y = |f(x)|$? In part (iii), start by assuming that $ff(4) = 0$ can be solved — and see where a problem arises.

Solutions

Question 1

$$y = (3x - 2)^4 \Rightarrow \frac{dy}{dx} = 12(3x - 2)^3$$

If gradient is 96, then $12(3x - 2)^3 = 96$

i.e. $(3x - 2)^3 = 8$, $3x - 2 = 2$, and hence $x = \frac{4}{3}$

When $x = \frac{4}{3}$, $y = (3 \times \frac{4}{3} - 2)^4 = 16$

So the point with gradient 96 is $(\frac{4}{3}, 16)$

 Not all candidates realise that differentiation is needed, and others carry out the differentiation incorrectly. Some candidates, having reached $12(3x - 2)^3 = 96$, proceed to expand and then have considerable difficulty solving the equation. However, for the majority of candidates, this question proves to be a successful start to the paper.

Question 2

(i) $1 \rightarrow 1.28772 \rightarrow 1.27571 \rightarrow 1.27609 \rightarrow 1.27607$

Hence $\alpha = 1.276$ (3 decimal places)

(ii) The equation is $x = (2 + e^{-2x})^{\frac{1}{3}}$ or $x^3 = 2 + e^{-2x}$

e Most candidates make effective use of a calculator to generate the terms of the sequence. It is necessary to show the terms — and not just the answer — to demonstrate to the examiner that the correct method has been used. The link between the iterative formula and the corresponding equation is not known by some candidates and thus they are unsure how to proceed in part (ii).

Question 3

Squaring both sides, $(3x - 5)^2 \leq (2x + 1)^2$

Expanding, $9x^2 - 30x + 25 \leq 4x^2 + 4x + 1$

So $5x^2 - 34x + 24 \leq 0$

Factorising, $(5x - 4)(x - 6) \leq 0$

Hence $\frac{4}{5} \leq x \leq 6$

e Different methods are possible, but the most popular is the one which involves squaring both sides to produce a quadratic inequality. However, many attempts to solve this inequality are poor and end with $(5x - 4)(x - 6) \leq 0$ giving $x \leq \frac{4}{5}, x \leq 6$. Reference to a sketch of $y = (5x - 4)(x - 6)$ to establish the correct set of values is strongly recommended.

Question 4

$$\text{Volume} = \int_{-1}^{19} \pi \frac{1}{2x+7} \, dx = \left[\frac{1}{2} \pi \ln|2x + 7| \right]_{-1}^{19}$$

$$= \frac{1}{2}\pi \ln 45 - \frac{1}{2}\pi \ln 5$$

$$= \frac{1}{2}\pi \ln \frac{45}{5} = \frac{1}{2}\pi \ln 9 = \pi \ln 3$$

e This question is testing the application of the formula for a volume of revolution as well as the use of logarithm properties to present the answer in a suitably simplified form. Most candidates answer this question competently, showing keen awareness of the relevant logarithm properties, thereby indicating performance of at least grade D standard.

Question 5

(i) From $\sin^2\theta + \cos^2\theta \equiv 1$ we have $\sin^2\theta + \frac{1}{9} = 1$ and hence $\sin^2\theta = \frac{8}{9}$

θ is acute so that $\sin\theta > 0$, giving $\sin\theta = \frac{2}{3}\sqrt{2}$

$\sin 2\theta = 2\sin\theta\cos\theta = 2 \times \frac{2}{3}\sqrt{2} \times \frac{1}{3} = \frac{4}{9}\sqrt{2}$

(ii) $\tan\theta = \dfrac{\sin\theta}{\cos\theta} = \frac{2}{3}\sqrt{2} \div \frac{1}{3} = 2\sqrt{2}$

$\tan(\theta - 45°) = \dfrac{\tan\theta - \tan 45°}{1 + \tan\theta\tan 45°} = \dfrac{2\sqrt{2} - 1}{1 + 2\sqrt{2} \times 1}$

$\qquad = \dfrac{(2\sqrt{2} - 1)(1 - 2\sqrt{2})}{(1 + 2\sqrt{2})(1 - 2\sqrt{2})} = \dfrac{2\sqrt{2} - 1 - 8 + 2\sqrt{2}}{1 - 8} = \dfrac{4\sqrt{2} - 9}{-7} = \frac{9}{7} - \frac{4}{7}\sqrt{2}$

 This question presents more problems to candidates; accurate and complete solutions to both parts indicate work of at least grade B standard. The key first step is to find the exact value of $\sin\theta$, and some candidates struggle with this. There is also uncertainty over the identity for $\sin 2\theta$. In part (ii), while the identity for expanding $\tan(\theta - 45°)$ is used accurately in general, subsequent work involving surds, using a technique met in unit C1, is often muddled.

Question 6

Volume $V\,m^3$ is given by $V = \frac{4}{3}\pi r^3$, which leads to $\dfrac{dV}{dr} = 4\pi r^2$, where r (in metres) is the radius

Given $\dfrac{dV}{dt} = 0.65$, to find $\dfrac{dr}{dt}$ when $r = 2$ (where t is time in minutes):

$\dfrac{dV}{dt} = \dfrac{dV}{dr} \times \dfrac{dr}{dt} \Rightarrow 0.65 = 4\pi \times 2^2 \times \dfrac{dr}{dt}$ at $r = 2$

so $\dfrac{dr}{dt} = \dfrac{0.65}{16\pi} = 0.013$

The radius is increasing at 0.013 metres per minute.

 Many candidates have a very limited understanding of connected rates of change, and part of the problem lies with a lack of appreciation of the distinction between derivatives such as $\dfrac{dV}{dr}$ and $\dfrac{dV}{dt}$. The question does not give the formula for the volume of a sphere and nor does the *List of Formulae* — this is a result that candidates are expected to know but which many, however, do not. A typical successful solution consists of accurate and concise statements involving the derivatives.

Question 7

(i) $y = x^2 e^{-3x} \Rightarrow y = 2xe^{-3x} - 3x^2 e^{-3x}$

For stationary points, $\dfrac{dy}{dx} = 0$, i.e. $2xe^{-3x} - 3x^2 e^{-3x} = 0$

Factorising gives $xe^{-3x}(2 - 3x) = 0$, leading to $x = 0$ or $x = \frac{2}{3}$ (since e^{-3x} can never be zero)

When $x = 0$, $y = 0$; this confirms that there is a stationary point at the origin.

When $x = \frac{2}{3}$, $y = \frac{4}{9}e^{-2}$

so P has coordinates $(\frac{2}{3}, \frac{4}{9}e^{-2})$

(ii) Area $\approx \frac{1}{3} \times \frac{1}{3}(0 + 4 \times \frac{1}{9}e^{-1} + \frac{4}{9}e^{-2}) \approx 0.0249$

🖉 Part (i) is answered well by most candidates who recognise the need to use the product rule. A few do not provide sufficient detail to confirm the stationary point at the origin, and some inappropriately give a decimal approximation to the y-coordinate of P. Candidates have more problems with part (ii), where many are perhaps unaccustomed to applying Simpson's rule with only two strips and cannot readily obtain $\frac{1}{3} h(y_0 + 4y_1 + y_2)$ from the Simpson's rule statement given in the *List of Formulae*.

Question 8

(i) $\tan^2\theta(\sec^2\theta + \text{cosec}^2\theta) \equiv \dfrac{\sin^2\theta}{\cos^2\theta}\left(\dfrac{1}{\cos^2\theta} + \dfrac{1}{\sin^2\theta}\right)$

$$\equiv \dfrac{\sin^2\theta\,(\sin^2\theta + \cos^2\theta)}{\cos^2\theta \times \sin^2\theta\cos^2\theta} \equiv \dfrac{\sin^2\theta \times 1}{\cos^4\theta\sin^2\theta} \equiv \dfrac{1}{\cos^4\theta} \equiv \sec^4\theta$$

(ii) (a) $5\tan^2\alpha(\sec^2\alpha + \text{cosec}^2\alpha) = 11$ is equivalent to $5\sec^4\alpha = 11$ by part (i)

This gives $\cos^4\alpha = \frac{5}{11}$, so $\cos\alpha = \pm\sqrt[4]{\frac{5}{11}}$

Hence $\alpha = 34.8°, 145.2°, 214.8°, 325.2°$

(ii) (b) $\tan^2\beta(\sec^2\beta + \text{cosec}^2\beta) = k$ is equivalent to $\sec^4\beta = k$

For all values of β, $\sec\beta \le -1$ or $\sec\beta \ge 1$ and therefore $\sec^4\beta \ge 1$

Thus, no values satisfy the equation if $k < 1$

🖉 This is a more challenging question, and many candidates have trouble constructing a clear and logical proof in part (i). Having tried a tentative first step, such as replacing $\tan^2\theta$ by $\sec^2\theta - 1$, but seeing no obvious way forward, candidates are often reluctant to abandon their initial attempt, doggedly persevering even when no progress towards the desired result is apparent. In part (ii), most candidates recognise the link with part (i), but it is very common in (a) for the two answers corresponding to $\sec\alpha = -\sqrt[4]{\frac{11}{5}}$ to be missed. Some awareness of the nature of the secant function is needed for (ii)(b). Candidates demonstrating that the value of k must be less than 1 are showing commendable mathematical ability.

Question 9

$y = 2e^{\frac{1}{4}x} \Rightarrow \dfrac{dy}{dx} = \frac{1}{2}e^{\frac{1}{4}x}$

When $x = \ln 16$, $\dfrac{dy}{dx} = \frac{1}{2}e^{\frac{1}{4}\ln 16} = \frac{1}{2}e^{\ln 2} = \frac{1}{2} \times 2 = 1$ and $y = 2e^{\frac{1}{4}\ln 16} = 4$

Gradient of the normal at P is -1

Equation of the normal at P is $y - 4 = -1(x - \ln 16)$, i.e. $y = -x + 4 + \ln 16$

The normal meets the x-axis where $y = 0$, giving $x = 4 + \ln 16$

Area under curve from y-axis to P is

$$\int_0^{\ln 16} 2e^{\frac{1}{4}x}\, dx = \left[8e^{\frac{1}{4}x}\right]_0^{\ln 16} = 8e^{\frac{1}{4}\ln 16} - 8e^0 = 16 - 8 = 8$$

Area of shaded region $= 8 + \frac{1}{2} \times 4 \times 4 = 16$

Although this question is unstructured, most candidates manage to establish the necessary steps to find the required area. The solution involves evaluation of $e^{\frac{1}{4}\ln 16}$ and the solution proceeds more easily if the value of this is quickly recognised as being 2. A common error occurs with the integration of $2e^{\frac{1}{4}x}$; a mental differentiation check would show that the integral is not $\frac{1}{2}e^{\frac{1}{4}x}$.

Question 10

(i) $f(x) = x^2 - 6x + c = (x - 3)^2 - 9 + c$, so the range is $f(x) \geq -9 + c$

Hence $-9 + c = 4$, giving $c = 13$

(ii) $f(x) = |f(x)|$ means that $f(x) \geq 0$ for all values of x

Hence $-9 + c \geq 0$ so that $c \geq 9$

(iii) $ff(4) = 0 \Rightarrow f(16 - 24 + c) = 0 \Rightarrow f(c - 8) = 0$

$$\Rightarrow (c - 8)^2 - 6(c - 8) + c = 0 \Rightarrow c^2 - 16c + 64 - 6c + 48 + c = 0$$

$$\Rightarrow c^2 - 21c + 112 = 0$$

Discriminant is $(-21)^2 - 4 \times 1 \times 112 = -7$

Since the discriminant is negative, the equation for c has no real roots, i.e. there are no values of c for which $ff(4) = 0$

Although the requests appear simple, the responses of many candidates reveal uncertainties about the theory involved; the advantage of expressing $f(x)$ in completed square form is not noted by many. In part (i), it is common for the answer to consist of a range of values, usually $c > 13$. A moment's thought in part (ii) about the nature of the graph of $y = f(x)$ is very helpful, but many candidates try instead to set up and solve an equation in x. The final part is more challenging, and candidates who proceed to show convincingly that there are no possible values of c are exhibiting work of grade A or B standard.

Specimen Paper 3

Question 1

(i) The curve $y = e^x$ can be transformed to the curve $y = e^{-x} - 2$ by a pair of transformations. Give details of these transformations. (3 marks)

(ii) Given that $e^{-x} - 2 = k$, express x in terms of k. (2 marks)

Question 2

Given that $f(x) = \ln(x^2 + 4)$, find $f''(x)$. (5 marks)

Question 3

(i) Given that $4\cos(\theta + 30°) = 3\cos(\theta - 30°)$, show that $\tan\theta = \frac{1}{7}\sqrt{3}$. (4 marks)

(ii) Hence solve, for $-180° < \theta < 180°$, the equation $4\cos(\theta + 30°) = 3\cos(\theta - 30°)$. (2 marks)

Question 4

Find the exact value of

(i) $\int_1^4 \dfrac{4}{(1 - 2x)^2}\,dx,$ (3 marks)

(ii) $\int_0^\infty 3e^{1-2x}\,dx.$ (3 marks)

Question 5

A particular substance is decaying exponentially. Its mass at a time t years from now is M grams. Its mass now is 180 grams, and ten years from now its mass will be 162 grams.

(i) Given that $M = Ae^{kt}$, find the values of the constants A and k. (3 marks)

(ii) In how many years from now will the mass be 90 grams? (2 marks)

(iii) Find the rate at which the mass is decreasing at a time 20 years from now. (3 marks)

Question 6

It is given that the equation $2x^3 + 7x - 24 = 0$ has exactly one real root.

(i) Determine by calculation the pair of consecutive integers between which the root lies. (3 marks)

(ii) Use the iterative formula $x_{n+1} = \sqrt[3]{12 - \frac{7}{2}x_n}$ with a suitable starting value to find the real root of $2x^3 + 7x - 24 = 0$ correct to 2 decimal places. (4 marks)

(iii) Find the real root of the equation $2e^{\frac{3}{2}x} + 7e^{\frac{1}{2}x} = 24$, giving the root correct to 2 decimal places. (2 marks)

Question 7

(i) Given that $y = (2x - 1)^5(3x + 1)^4$, show that $\dfrac{dy}{dx}$ can be expressed in the form

$(2x - 1)^4(3x + 1)^3(ax + b)$, where the constants a and b are to be determined. (5 marks)

(ii) Hence find the equation of the tangent to the curve $y = (2x - 1)^5(3x + 1)^4$ at the point where it crosses the y-axis. (3 marks)

Question 8

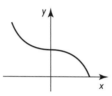

The diagram shows the curve $y = f(x)$, where the function f is defined by $f(x) = \cos^{-1} x$ with domain $-1 \le x \le 1$.

(i) State the range of f. (1 mark)

(ii) Evaluate $ff(\tfrac{1}{2}\sqrt{3})$ correct to 3 significant figures. (3 marks)

(iii) Sketch the curve $y = f^{-1}(x)$. (2 marks)

Question 9

(i) Prove that $\dfrac{\sin^2 2\theta}{1 + \cos 2\theta} \equiv 2\sin^2\theta$. (4 marks)

(ii) Hence show that $\sin\tfrac{1}{12}\pi = \tfrac{1}{2}\sqrt{2 - \sqrt{3}}$. (4 marks)

Question 10

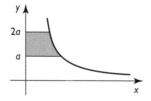

The diagram shows part of the curve $y = \dfrac{4}{x - 1}$. The shaded region is bounded by the curve and the lines $x = 0$, $y = a$ and $y = 2a$, where a is a positive constant. The shaded region is rotated completely about the y-axis. Given that the volume of the solid produced is 100 cubic units, find the possible values of a correct to 3 significant figures. (11 marks)

Solutions

Question 1

(i) Reflection in the y-axis and translation in the negative y-direction by 2 units (in either order).

(ii) $e^{-x} - 2 = k \Rightarrow e^{-x} = k + 2 \Rightarrow -x = \ln(k + 2)$, so $x = -\ln(k + 2)$

📝 The vast majority of candidates recognise that a reflection and a translation are involved, but not all provide accurate details of the two transformations. Part (ii) is answered well by most candidates, apart from a few whose first step is the unfortunate $-x - \ln 2 = \ln k$.

Question 2

$f(x) = \ln(x^2 + 4) \Rightarrow f'(x) = \dfrac{2x}{x^2 + 4} \Rightarrow f''(x) = \dfrac{(x^2 + 4) \cdot 2 - 2x \cdot 2x}{(x^2 + 4)^2} = \dfrac{8 - 2x^2}{(x^2 + 4)^2}$

📝 Most candidates find the derivatives accurately, first using the chain rule and then realising that the quotient rule is needed to obtain the second derivative. The error of differentiating $\ln(x^2 + 4)$ to give $\dfrac{1}{x^2 + 4}$ is not common.

Question 3

(i) Expanding both sides gives

$4(\cos\theta\cos 30° - \sin\theta\sin 30°) = 3(\cos\theta\cos 30° + \sin\theta\sin 30°)$

which leads to $2\sqrt{3}\cos\theta - 2\sin\theta = \frac{3}{2}\sqrt{3}\cos\theta + \frac{3}{2}\sin\theta$

and hence $\frac{1}{2}\sqrt{3}\cos\theta = \frac{7}{2}\sin\theta$

Dividing by $\cos\theta$, we obtain $\frac{1}{2}\sqrt{3} = \frac{7}{2}\tan\theta$, so $\tan\theta = \frac{1}{7}\sqrt{3}$

(ii) $\tan\theta = \frac{1}{7}\sqrt{3}$ and $-180° < \theta < 180° \Rightarrow \theta = 13.9°, -166.1°$

📝 Some candidates are unsure how to begin and do not make use of the angle-sum identities given in the *List of Formulae* booklet. For those candidates who do expand, the step of dividing through by $\cos\theta$ to produce $\tan\theta$ eludes some. Because candidates are often asked to provide angles in the range 0° to 360°, the different request here means that the angle $-166.1°$ is missed by some.

Question 4

(i) $\displaystyle\int_1^4 \frac{4}{(1-2x)^2}\,dx = \int_1^4 4(1-2x)^{-2}\,dx = \left[\frac{4}{(-1)(-2)}(1-2x)^{-1}\right]_1^4 = \left[\frac{2}{1-2x}\right]_1^4 = -\frac{2}{7} - (-2) = \frac{12}{7}$

(ii) $\displaystyle\int_0^\infty 3e^{1-2x}\,dx = \left[-\frac{3}{2}e^{1-2x}\right]_0^\infty = 0 - (-\frac{3}{2}e^1) = \frac{3}{2}e$

Most candidates answer part (i) correctly, although some make a mistake with the factor of $(1 - 2x)^{-1}$. Part (ii) presents more problems because of the upper limit ∞. Many candidates provide an answer involving $e^{1-2\infty}$; an appreciation that the value of e^{1-2x} tends to zero as x tends to infinity is often not apparent. Two completely correct solutions indicate work of at least grade B standard.

Question 5

(i) Substituting $t = 0$ and $M = 180$ into $M = Ae^{kt}$ gives $180 = Ae^0$ and hence $A = 180$

Substituting $t = 10$ and $M = 162$ gives $162 = 180e^{10k}$

So $e^{10k} = 0.9$, $10k = \ln 0.9$

and hence $k = \frac{1}{10}\ln 0.9 = -0.010536$ (to 5 significant figures)

(ii) $90 = 180e^{(\frac{1}{10}\ln 0.9)t} \Rightarrow e^{(\frac{1}{10}\ln 0.9)t} = 0.5 \Rightarrow (\frac{1}{10}\ln 0.9)t = \ln 0.5 \Rightarrow t = 65.8$

Thus, the mass will be $90\,g$ in 65.8 years from now.

(iii) $M = 180e^{(\frac{1}{10}\ln 0.9)t} \Rightarrow \dfrac{dM}{dt} = 180 \times (\frac{1}{10}\ln 0.9)e^{(\frac{1}{10}\ln 0.9)t}$

Substituting $t = 20$, $\dfrac{dM}{dt} = 18(\ln 0.9)e^{2\ln 0.9} = -1.54$

So, 20 years from now, the mass is decreasing at 1.54 grams per year.

 This question is handled with assurance by most candidates and, generally, all three parts are answered competently. The value of k is $\frac{1}{10}\ln 0.9$; this exact value may be used in the calculations of parts (ii) and (iii) as can an approximate value such as -0.010536. But be wary of using too approximate a value, like -0.011 (with only two significant figures), because this will lead to inaccurate answers for later parts. Candidates should carry out the working with an accurate value of k and then present final answers to a degree of accuracy appropriate to the context of the question; giving answers to 2 or 3 significant figures is acceptable in this case.

Question 6

(i) Let $f(x) = 2x^3 + 7x - 24$

Substituting integer values, $f(0) = -24$, $f(1) = -15$, $f(2) = 6$

The change of sign means that the root is between 1 and 2

(ii) Starting with $x_1 = 1.5$, the iterative formula gives

$x_2 = 1.88988$, $x_3 = 1.75283$, $x_4 = 1.80340$, $x_5 = 1.78507$, $x_6 = 1.79176$, $x_7 = 1.78932$

The root is 1.79 (correct to 2 d.p.)

(iii) Substituting $u = e^{\frac{1}{2}x}$ gives $2u^3 + 7u = 24$, i.e. $2u^3 + 7u - 24 = 0$

Hence, from part (ii), $u = 1.78932$

Therefore $e^{\frac{1}{2}x} = 1.78932$, so $\frac{1}{2}x = \ln 1.78932$ and hence $x = 2\ln 1.78932 = 1.16$ (correct to 2 d.p.)

e The first two parts are answered well, with candidates generally providing suitable evidence of their methods. Part (iii) is more challenging, and not all candidates recognise how it is linked with the earlier parts.

Question 7

(i) $y = (2x - 1)^5(3x + 1)^4 \Rightarrow \dfrac{dy}{dx} = 10(2x - 1)^4(3x + 1)^4 + 12(2x - 1)^5(3x + 1)^3$

$$= (2x - 1)^4(3x + 1)^3[10(3x + 1) + 12(2x - 1)]$$

$$= (2x - 1)^4(3x + 1)^3(54x - 2)$$

(ii) The curve crosses the y-axis when $x = 0$

At $x = 0, y = -1$ and $\dfrac{dy}{dx} = -2$

So the tangent has equation $y - (-1) = -2(x - 0)$, i.e. $y = -2x - 1$ or $2x + y + 1 = 0$

e Candidates know to use the product rule for the differentiation but not all manage to apply the chain rule correctly. Presenting the derivative in a specified form is a less familiar request; many candidates do not notice that $(2x - 1)^4(3x + 1)^3$ is a common factor and make matters far too complicated by attempting a complete expansion of the expression for the derivative. Part (ii) tends to be answered well.

Question 8

(i) Range of f is $0 \le f(x) \le \pi$

(ii) $ff(\tfrac{1}{2}\sqrt{3}) = f(\cos^{-1}(\tfrac{1}{2}\sqrt{3})) = f(\tfrac{1}{6}\pi) = \cos^{-1}(\tfrac{1}{6}\pi) = 1.02$ (to 3 s.f.)

(iii) This is the graph of $\cos x$ for $0 \le x \le \pi$

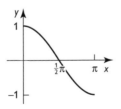

e Many candidates are not so confident in answering questions involving inverse trigonometric functions. Although this question is reasonably straightforward, many candidates have trouble with it. Not all are sure what is meant by 'range', and not using radians is a common mistake in part (ii). In part (iii), some candidates realise that $f^{-1}(x) = \cos x$ but offer a sketch for a domain greater than $0 \le x \le \pi$.

Question 9

(i) $\dfrac{\sin^2 2\theta}{1 + \cos 2\theta} \equiv \dfrac{(2\sin\theta\cos\theta)^2}{1 + \cos^2\theta - \sin^2\theta}$

$$\equiv \dfrac{4\sin^2\theta\cos^2\theta}{\cos^2\theta + (1 - \sin^2\theta)} \equiv \dfrac{4\sin^2\theta\cos^2\theta}{2\cos^2\theta} \equiv 2\sin^2\theta$$

(ii) Putting $\theta = \frac{1}{12}\pi$, we get $2\sin^2\frac{1}{12}\pi = \dfrac{\sin^2\frac{1}{6}\pi}{1+\cos\frac{1}{6}\pi} = \dfrac{\frac{1}{4}}{1+\frac{1}{2}\sqrt{3}} = \dfrac{1}{4+2\sqrt{3}}$

Rationalising the denominator, $2\sin^2\frac{1}{12}\pi = \dfrac{4-2\sqrt{3}}{(4+2\sqrt{3})(4-2\sqrt{3})} = \dfrac{4-2\sqrt{3}}{16-12} = \frac{1}{2}(2-\sqrt{3})$

Hence $\sin^2\frac{1}{12}\pi = \frac{1}{4}(2-\sqrt{3})$ and so $\sin\frac{1}{12}\pi = \frac{1}{2}\sqrt{2-\sqrt{3}}$, where we have chosen the positive root because $\frac{1}{12}\pi$ is a value in the first quadrant.

 The proof in part (i) requires the double-angle formulae and not all candidates are confident in using these. The double-angle identities are needed sufficiently often in this module that candidates are advised to memorise the formulae for $\sin 2\theta$, $\cos 2\theta$ and $\tan 2\theta$. Part (ii) depends on recalling material from module C1, but many candidates are unable to remember the exact value of $\cos\frac{1}{6}\pi$ or the process for rationalising a denominator.

Question 10

$y = \dfrac{4}{x-1} \Rightarrow x - 1 = \dfrac{4}{y} \Rightarrow x = 1 + \dfrac{4}{y}$

Volume of solid $= \displaystyle\int_a^{2a} \pi(1+\frac{4}{y})^2 dy$

$\displaystyle = \int_a^{2a} \pi(1+\frac{8}{y}+\frac{16}{y^2})dy = \pi\left[y + 8\ln y - \frac{16}{y}\right]_a^{2a}$

$= \pi(2a + 8\ln 2a - \dfrac{16}{2a}) - \pi(a + 8\ln a - \dfrac{16}{a})$

$= \pi(a + 8(\ln 2a - \ln a) + \dfrac{8}{a}) = \pi(a + 8\ln 2 + \dfrac{8}{a})$

Volume is 100, so $\pi(a + 8\ln 2 + \dfrac{8}{a}) = 100$

Hence $\pi a^2 + (8\pi\ln 2 - 100)a + 8\pi = 0$

giving $a = \dfrac{-(8\pi\ln 2 - 100) \pm \sqrt{(8\pi\ln 2 - 100)^2 - 4 \cdot \pi \cdot 8\pi}}{2\pi} = 0.308$ or 26.0

 Candidates are very familiar with finding a volume when a region is rotated about the x-axis but tend to be less assured when rotation about the y-axis is involved. Some are unaware of the formula $\int_c^d \pi x^2 dy$, and the rearrangement of $y = \dfrac{4}{x-1}$ is not always carried out competently. Simplification of $\ln 2a - \ln a$ to $\ln 2$ is a crucial step after the integration has been performed. To find the possible values of a, the solution of a complicated quadratic equation is called for and great care is needed. A correct outcome to this question indicates work of a very high standard.

Answers to exercises

Exercise 1

(1) $f(x) \geq -5$

(2) $-1 \leq g(x) \leq 1$

(3) $2x^2 + 4x + 11 = 2(x + 1)^2 + 9$, so $f(x) \geq 9$

Exercise 2

(1) (i) 23   **(ii)** $14x - 4$

(2) (i) $2\sqrt{6}$ **(ii)** $2^{\sqrt{x-8}}$

Exercise 3

(1) $\sqrt[3]{2x - 10}$

(2) 9

(3)

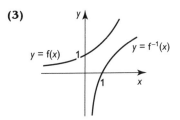

Exercise 4

(1)

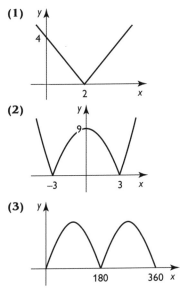

(2)

(3)

Exercise 5

(1) –8, 3

(2) –2

(3) –5, $\frac{9}{5}$

Exercise 6

(1) $x < -\frac{1}{2}$ or $x > 1$

(2) $x < 1$

(3) $x \leq -1$ or $x \geq 1$

Exercise 7

(1) $y = \frac{1}{2}|x - 4|$

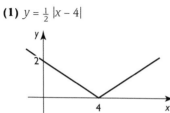

(2) Translation by 5 units in the negative x-direction, followed by a stretch in the x-direction with scale factor $\frac{1}{2}$ (the order of applying these two steps is important), together with reflection in the x-axis (which can be applied at any stage)

Exercise 8

(1) $e^{\frac{3}{2}} - 5$

(2) $\ln(\frac{8}{3}e^5)$

(3) 0.886

Exercise 9

(1) $f^{-1}(x) = \frac{1}{3}(\log_2 x + 1)$ or $\frac{1}{3}\left(\frac{\ln x}{\ln 2} + 1\right)$

(2) $x = -8a$ or $x = -a$

(3) $f(x) = 16 - 4(x - \frac{1}{2})^2$; range $f(x) \leq 16$

(4) $\frac{6}{5} < x < 8$

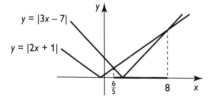

(5) Graph:

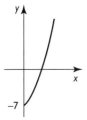

Any horizontal line $y = k$ with $k \geq -7$ meets the curve exactly once, so f is 1–1

$f^{-1}(x) = \frac{1}{2}(\sqrt{2x+15} - 1)$

Exercise 10

(1) $12x^3 + x^{-2}$

(2) $\frac{1}{6}$

(3) $y = -3x - 17$

(4) $8x^{-3}$

(5) $(-6, 227)$ maximum; $(2, -29)$ minimum

Exercise 11

(1) $y = -10e^{-5x} - 2e^{\frac{1}{2}x}$

(2) 12

(3) $y = 13x + 8$

(4) $-\frac{1}{4}\ln 2$

Exercise 12

(1) $3x^2 + 1 - \frac{8}{x}$

(2) $(\frac{1}{4}, -3\ln 4)$

(3) 3

Exercise 13

(1) $40(4x - 5)^9$

(2) $\frac{2x}{x^2 + 4}$

(3) $18e^{3x}(e^{3x} + 2)^5$

(4) $-\frac{4(2x+1)}{(x^2 + x + 1)^2}$

(5) $-16(8x + 1)^{-\frac{3}{2}}$

Exercise 14

(1) $3x^2 \ln x + x^2$

(2) $2x(3x+1)^{\frac{1}{2}} + \frac{3}{2}x^2(3x+1)^{-\frac{1}{2}}$

(3) $y = 12x - 11$

Exercise 15

(1) $\dfrac{4 - 4x^2}{(x^2 + 1)^2}$

(2) $\dfrac{2}{e(e + 2)^2}$

(3) $(-1, -\frac{1}{2})$ and $(3, \frac{1}{6})$

Exercise 16

(1) 38.5 years

(2) 22

(3) 270, 405, 1367

Exercise 17

(1) $0.01 \, \text{cm s}^{-1}$

(2) $101 \, \text{m}^2$ per day

Exercise 18

(1) $(18, 375\sqrt{3})$

(2) $0, \dfrac{2}{3}$

(3) $5x - 2y - 12 = 0$

(4) (i) 3840   **(ii)** $120e^{\left(\frac{1}{12}\ln 2\right)t}$ or $120 \times 2^{\frac{1}{12}t}$ **(iii)** 114

Exercise 19

(1) $2x^4 + x + c$

(2) $6x^{\frac{1}{2}} - \frac{1}{2}x^2 + c$

(3) $72\frac{2}{3}$

(4) 5

Exercise 20

(1) $\frac{1}{2}(2x - 3)^{10} + c$

(2) $\frac{1}{4}\sqrt{8x + 3} + c$

(3) $\frac{4}{15}$

Exercise 21

(1) $\ln 343$

(2) $\ln 3$

Exercise 22

(1) $2e^{3x} - 2e^{-2x} + c$

(2) $5(e^2 - 1)$

Exercise 23

(1) 18π

(2) $\frac{1}{18}\pi$

(3) $\frac{56}{15}\pi$

Exercise 24

(1) 20.6

(2) 2.61

Exercise 25

(1) 37

(2) $\frac{11}{6}$

(3) 26.6

(4) $\frac{1}{4}\pi(e^{12} + 4e^6 + 7)$

(5) $\frac{125}{18}$

Exercise 26

(1) $3 + \sin^2\theta$

(2) $66.0°, 246.0°$

(3) $48.2°, 120°, 240°, 311.8°$

Exercise 27

(1) $19.2°, 340.8°$

(2) $41.8°, 90°, 138.2°, 270°$

(3) $60°, 75.5°, 284.5°, 300°$

Exercise 28

(1) $\frac{3}{13}\sqrt{3}$

(3) $\frac{1}{2}\sqrt{2}+\frac{1}{6}\sqrt{3}$

Exercise 29

(1) $17\cos(\theta - 61.9°)$

(2) $\sqrt{2}\sin(\theta + 45°)$

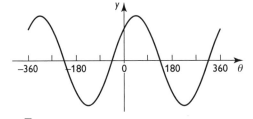

(3) $3\sqrt{5}\cos(\theta + 63.4°)$; $9.2°$, $223.9°$

Exercise 30

(1) $-\dfrac{7}{18}$

(2) 0.588, 2.554, 3.730, 5.695

(3) $61.0°$, $90°$, $119.0°$, $270°$

Exercise 31

(1) $\frac{1}{6}\pi$

(2) $-\frac{1}{4}\pi$

(3) $\frac{2}{3}\pi$

Exercise 32

(1) $23.6°$, $156.4°$, $194.5°$, $345.5°$

(2) (i) $\dfrac{240}{289}$

   (ii) $\frac{1}{34}(15-8\sqrt{3})$

(3) $\sqrt{7}\cos(\theta + 32.3°)$; $-73.2°$, $8.6°$

(4) (ii) $\dfrac{\sqrt{3}+1}{2\sqrt{2}}$

(6) $-\dfrac{24}{7}$

Exercise 33

(1) (i) 2   **(ii)** 1   **(iii)** 3

Exercise 34

(1) 1 and 2

Exercise 35

(1) (iii) 1.4082

(2) (ii) $-1 < \alpha < 0, 3 < \beta < 4$ **(iii)** $\alpha = -0.9096, \beta = 3.8171$

Exercise 36

(1) 3.317, 3.411, 3.438, 3.446, 3.449

(2) 0.270, 2.872

(3) 14.9°, 194.9°

(4)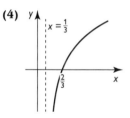

(5) $4 \ln 3$

Exercise 37

(1) $2 \sin A \cos A$

(2) $\dfrac{a}{ax+b}$

(3) $b^2 - 4ac < 0$

(4) $\frac{2}{\sqrt{3}}$ or, equivalently, $\frac{2}{3}\sqrt{3}$

(5) $1 + \tan^2 \theta \equiv \sec^2 \theta$

(6) ke^{kx}

(7) $\int_a^b \pi y^2 dx$

(8) $\ln \dfrac{a^2 b}{c}$

(9) $\dfrac{2 \tan A}{1 - \tan^2 A}$

(10) If $y = uv$, then $\dfrac{dy}{dx} = \dfrac{du}{dx} \cdot v + u \cdot \dfrac{dv}{dx}$

(11) $b^2 - 4ac$

(12) $-\sqrt{3}$

(13) $1 + \cot^2\theta \equiv \mathrm{cosec}^2\theta$

(14) $\int_c^d \pi x^2 \mathrm{d}y$

(15) $\frac{1}{k}e^{kx} + c$

(16) $2\cos^2\theta - 1$

(17) $-\sqrt{2}$

(18) $\frac{1}{a}\ln|ax + b| + c$

(19) $t = 10^{Q/2} - 1$

(20)

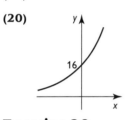

Exercise 38

(1) $2\ln\frac{3}{2}$

(2) $x - 2y + 7 = 0$

(3) $f(x) \geq -5\frac{1}{2}$

(5) $\frac{1}{7} < x < \frac{1}{3}$

(6) $y = 20 - 5(3 - 2x)^8$

(7) $\ln 10$